8th International Workshop on Multi-Rate Processes and Hysteresis and the 3rd International Workshop on Hysteresis and Slow-Fast Systems (MURPHYS-HSFS 2016)

Journal of Physics: Conference Series Volume 811

Barcelona, Spain
13 – 17 June 2016

Editors:

Pavel Gurevich
Andrei Korobeinikov
Dmitrii Rachinskii
Vladimir Sobolev

ISBN: 978-1-5108-3719-5
ISSN: 1742-6588

Printed from e-media with permission by:

Curran Associates, Inc.
57 Morehouse Lane
Red Hook, NY 12571

Printed by Curran Associates, Inc. (2017)

For permission requests, please contact the Institute of Physics
at the address below.

Institute of Physics
Dirac House, Temple Back
Bristol BS1 6BE UK

Phone: 44 1 17 929 7481
Fax: 44 1 17 920 0979

techtracking@iop.org

Additional copies of this publication are available from:

Curran Associates, Inc.
57 Morehouse Lane
Red Hook, NY 12571 USA
Phone: 845-758-0400
Fax: 845-758-2633
Email: curran@proceedings.com
Web: www.proceedings.com

Table of contents

Volume 811

8th Workshop on Multi-Rate Processes and Hysteresis and the HSFS Workshop (Hysteresis and Slow-Fast Systems)

13–17 June 2016, Barcelona, Spain

Accepted papers received: 14 February 2017
Published online: 3 March 2017

Preface

Papers

Foreword

The current issue of *Journal of Physics: Conference Series* offers to your attention a selection of papers presented at the joint 8th International Workshop on **MU**lti-**R**ate **P**rocesses and **HYS**teresis (MURPHYS) and 3rd International Workshop on **H**ysteresis and **S**low-**F**ast **S**ystems (HSFS) that was hosted by the Centre de Recerca Matemàtica (www.crm.cat), Barcelona in June 13 to 17 of 2016. This meeting, MURPHYS-HSFS-2016, continued a successful series of biennial multidisciplinary conferences on Multi-Rate Processes and Hysteresis, that previously took place in Cork (Ireland, 2002 to 2008), Pécs (Hungary, 2010), Suceava (Romania, 2012) and Berlin (Germany, 2014), as well as the series of workshops on Hysteresis and Slow-Fast Systems in Lutherstadt, Wittenberg and Berlin.

MURPHYS-HSFS-2016 workshop brought together over 30 researchers working on hysteresis and multi-scale phenomena from Europe, Russia and US. Participants shared and discussed recent developments of analytical techniques in several areas of common interest. Topics of this volume include analysis of hysteresis phenomena, multiple scale systems, self-organizing nonlinear systems, singular perturbations and critical phenomena, as well as applications of the hysteresis and the theory of singularly perturbed systems to fluid dynamics, chemical kinetics, cancer modeling, population modeling, mathematical economics and control. More information about the Workshop can be found at:

http://www.crm.cat/en/Activities/Curs_2015-2016/Pages/MURPHYS.aspx

All papers published in this volume of *Journal of Physics: Conference Series* have been peer reviewed through processes administered by the Editors. Reviews were conducted by expert referees to the professional and scientific standards expected of a proceedings journal published by IOP Publishing.

We are happy to acknowledge support of the Workshop by AGAUR of the Generalitat de Catalunya, CERCA Programme of the Generalitat de Catalunya, the Centre de Recerca Matemàtica, Collaborative Research Center 910: Control of Self-organizing Nonlinear Systems (Germany), the Samara National Research University (Russia) and Drexel University (USA).

Pavel Gurevich, Andrei Korobeinikov, Dmitrii Rachinskii and Vladimir Sobolev,
Editors

IOP Conf. Series: Journal of Physics: Conf. Series **811** (2017) 011002 doi:10.1088/1742-6596/811/1/011002

Peer review statement

All papers published in this volume of *Journal of Physics: Conference Series* have been peer reviewed through processes administered by the proceedings Editors. Reviews were conducted by expert referees to the professional and scientific standards expected of a proceedings journal published by IOP Publishing.

Slow and fast diffusion in a lead sulphate gravity separation process

Vincent Cregan[1,a], **William T. Lee**[2,b]

[1]Centre de Recerca Matemàtica, Campus de Bellaterra, Edifici C, 08193 Bellaterra, Barcelona, Spain

[2] Deptartment of Mathematics, University of Portsmouth, Winston Churchill Ave, Portsmouth PO1 2UP, UK

E-mail: [a]vcregan@crm.cat, [b]william.lee@port.ac.uk

Abstract. A model for the growth of lead sulphate particles in a gravity separation system from the crystal glassware industry is presented. The lead sulphate particles are an undesirable byproduct, and thus the model is used to ascertain the optimal system temperature configuration such that particle extraction is maximised. The model describes the evolution of a single, spherical particle due to the mass flux of lead particles from a surrounding acid solution. We divide the concentration field into two separate regions. Specifically, a relatively small boundary layer region around the particle is characterised by fast diffusion, and is thus considered quasi-static. In contrast, diffusion in the far-field is slower, and hence assumed to be time-dependent. The final system consisting of two nonlinear, coupled ordinary differential equations for the particle radius and lead concentration, is integrated numerically.

1. Introduction

Particle settling (or gravity separation) is a technique used in industry to separate liquid suspensions into one or more of its constitutive solid components by the action of gravity. Typical gravity separation industrial and medical applications include the removal of silt from industrial waste [1], the extraction of coal from material matter [2], gold refinery [3] and the division of blood into its fundamental components [4].

In the present study we consider a gravity settling method used to extract lead sulphate particles in acid polishing of lead crystal glassware. This process is currently used by an Irish glassware manufacturer and was originally presented at the 62nd European Study Group with Industry (ESGI62) [5]. The process begins with grinding where design cuts are inscribed onto the blank surfaces of the glass leaving it damaged and optically opaque. The damaged glass is then immersed in a polishing tank containing an aqueous solution of 65% sulphuric acid and 3% hydrofluoric acid. Chemical reactions between the acids and glass result in the formation of an insoluble lead sulphate layer at the glass surface. This layer obstructs the solution from reaching the glass, and hence the polishing performance deteriorates over time. Growth of the layer is counteracted by rinsing the glass regularly in an acidic-water solution. After rinsing, the solution is pumped into a large cone-shaped settlement tank where the lead sulphate particles settle out. The acid solution is cooled in the settlement tank such that the lead sulphate concentration decreases below the corresponding solubility limit in the polishing tank. This decrease in solubility results in both the precipitation of new small lead sulphate particles and the growth of existing particles. However, the formation of the new particles is not desirable as they take longer to settle compared to existing particles. In addition, if the crystals are not given enough time to settle they are reintroduced back into the warmer polishing tank where they dissolve and bring the acid solution closer to saturation. After a period of time the acid solution once again becomes saturated and the insoluble lead sulphate layer reforms on the glassware. Repeated use of the settlement tank decreases the time for which the glassware can be polished effectively.

Previous particle settling studies have generally focussed on the dimensions of the settlement tank [1, 6], the fluid flow regime [7] or the introduction of an appropriate chemical agent [8]. The aim of this work is to investigate the often neglected influence that temperature has on gravity separation. In particular, we consider the effect of different temperature configurations on the growth of insoluble lead sulphate particles. We formulate a model for the evolution of a single, spherical lead sulphate particle due to the mass flux of lead atoms from the surrounding acid solution. We simplify the model to a system of two nonlinear ordinary differential equations for the lead concentration in the acid solution and the particle radius. We use this system to investigate the effect that three different temperature configurations have on process performance. Initially, we consider the current constant temperature regime where the initial temperature of the acid solution adjusts to the tank temperature. Following this we study the effect of cooling the acid solution at slower rates. Numerical solutions for the lead concentration and particle radius for each of the temperature regimes are presented. Via homogeneous nucleation theory, we provide a criterion which determines growth of existing lead sulphate particles and the dissolution of new, smaller particles.

The outline of this paper is as follows. In Section 2 a model for the growth of a single, spherical lead sulphate particle is presented. The various temperature configurations are also highlighted. The governing concentration and particle radius equations are introduced in Section 3. The model for each of the temperature configurations is nondimensionalised in Section 4. The numerical results are discussed in Section 5.

2. Mathematical model

We consider a single, spherical lead sulphate particle with radius $r_p \equiv r_p(t)$ growing with time t in the acid solution. We apply a similar modelling approach used by Liger-Belair *et al.* [9] for the case of carbon dioxide bubble nucleation in champagne whereby we assume that the spherical domain around the particle nucleus can be approximated by two spherical subdomains: a relatively small boundary layer region around the particle (referred to as the diffusion region) and a larger far field well-mixed region (see Figure 1). The diffusion and well-mixed regions are characterised by time-independent and time-dependent lead concentration regimes, respectively. The radii of the diffusion and well-mixed regions are denoted r_d and r_w, respectively, such that $r_w \gg r_d$. We assume that the solution is sufficiently dilute such that particles interactions are negligible.

Assuming that the solution has an initial, constant lead concentration, C_0, the governing equations are

$$\frac{D}{r^2}\frac{\mathrm{d}}{\mathrm{d}r}\left(r^2\frac{\mathrm{d}C_d}{\mathrm{d}r}\right) = 0\,, \qquad \frac{\mathrm{d}C_w}{\mathrm{d}t} = 0\,, \tag{1}$$

where C_d and C_w are the lead concentrations in the diffusion and well-mixed regions, respectively, r is the radial coordinate, t is time and D is the diffusion coefficient for the lead atoms. At the interface of the diffusion region and the well-mixed region we prescribe

$$C_d(r_d) = C_w(r_d)\,, \qquad V_w\frac{\mathrm{d}C_w}{\mathrm{d}t} = -4\pi r_d^2 D\left.\frac{\mathrm{d}C_d}{\mathrm{d}r}\right|_{r=r_d}\,, \tag{2}$$

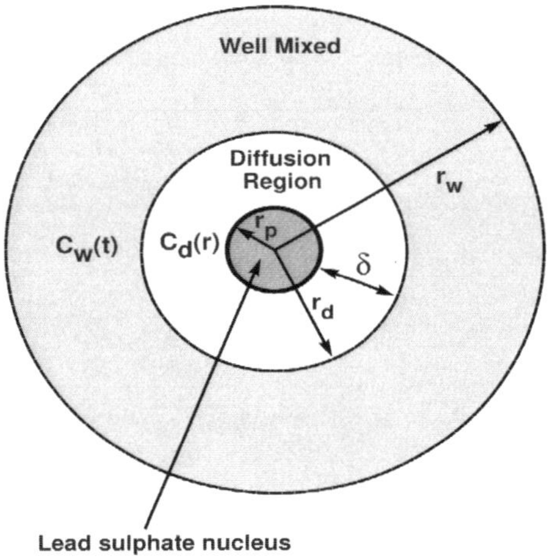

Figure 1. Lead sulphate particle with surrounding diffusion and well-mixed regions.

where V_w is volume of the well-mixed region. The first of these conditions represents lead concentration continuity condition, whilst the latter accounts for lead atoms diffusing between both regions. To track the surface of the particle we use the Stefan-type condition

$$\frac{\mathrm{d}r_p}{\mathrm{d}t} = DV_M \left.\frac{\partial C_d}{\partial r}\right|_{r=r_p}, \tag{3}$$

where V_M is the molar volume. The initial conditions are

$$r_p(0) = R_0, \qquad C_w(0) = C_0, \tag{4}$$

where R_0 is the initial particle radius.

The particle solubility is given by the Ostwald-Freundlich condition

$$C_d|_{r=r_p} = C^{EQ}(T)\exp\left(\frac{2\sigma V_M}{R_G T r_p}\right), \tag{5}$$

where σ is surface tension, R_G is the universal gas constant and T is tank temperature.

From [10], the particle critical radius is

$$r_c = \frac{2\sigma V_M}{R_G T \ln\left(C_w/C^{EQ}\right)}, \tag{6}$$

where the particle grows if $r_p > r_c$ and dissolves if $r_p < r_c$. Hence, r_c acts as a criterion for whether particles will dissolve or grow, and hence can be used to ascertain the effectiveness of the particle settling process.

We consider three different temperature regimes, namely a constant temperature regime and two time-dependent temperature configurations. The process's current temperature configuration is defined by two temperatures: the initial temperature of the acid solution as it enters the tank, T_0 and the final steady state temperature of the solution, T_1. Whilst the initial temperature is fixed, the final temperature is adjustable. We note that for particle growth, we require $T_1 < T_0$. To model the present operating set-up, we assume that the acid solution

IOP Conf. Series: Journal of Physics: Conf. Series **811** (2017) 012001 doi:10.1088/1742-6596/811/1/012001

changes rapidly from T_0 to T_1. To investigate the effect of cooling the acid solution at slower rates than that of the current set-up, we define two monotonically decreasing time-dependent temperature functions

$$T_L = \begin{cases} T_0 + (T_1 - T_0)t/\tau & \text{if} \quad t \leq \tau \\ T_1 & \text{if} \quad t \geq \tau\,, \end{cases} \tag{7}$$

$$T_E = T_1 - (T_1 - T_0)\,\mathrm{erfc}\left(\frac{t}{\tau}\right), \tag{8}$$

where τ is a characteristic lead diffusion time scale associated with the constant temperature problem and erfc is the complementary error function. For the proceeding analysis the C, L and E subscripts denote terms pertaining to the constant, linear and error temperature models, respectively. Functions (7) and (8) are designed such that at time $t = 0$ (when the acid solution enters the tank) the temperature is equal to T_0. Thereafter the temperature decreases with time until the temperature is exactly equal to T_1 at $t = \tau$ for (7) and approximately equal to T_1 at $t = \tau$ for (8).

3. Solution

Solving the first equation of (1) leads to

$$C_d(r) = \frac{1}{r(r_p - r_d)}\left(r_d(r_p - r)C_w - C^{EQ}\exp\left(\frac{2\sigma V_M}{R_G T r_p}\right) r_p(r_d - r)\right), \tag{9}$$

where the concentration continuity condition in (2) and the solubility condition in (5) were used. Introducing (9) into the second interface condition in (2) yields

$$\frac{\mathrm{d}C_w}{\mathrm{d}t} = -\frac{3\,r_d\,r_p\,D}{r_w^3\,(r_p - r_d)}\left(C^{EQ}\,\exp\left(\frac{2\sigma V_M}{R_G T r_p}\right) - C_w\right), \tag{10}$$

where $V_W \equiv 4\pi r_w^3/3$ has been exploited. The particle radius equation is obtained by substituting (9) into the Stefan condition (3) to give

$$\frac{\mathrm{d}r_p}{\mathrm{d}t} = \frac{D V_M r_d}{r_p\,(r_p - r_d)}\left(C^{EQ}\,\exp\left(\frac{2\sigma V_M}{R_G T r_p}\right) - C_w\right). \tag{11}$$

4. Nondimensionalisation

4.1. Constant temperature model

Rewriting (10) and (11) in terms of the tank temperature T_1 and exploiting $\delta = r_d - r_p$ leads to

$$\frac{\mathrm{d}C_w}{\mathrm{d}t} = \frac{3\,D}{\delta r_w^3}\,r_p\,(r_p + \delta)\left(C^{EQ}(T_1)\,\exp\left(\frac{2\sigma V_M}{R_G T_1 r_p}\right) - C_w\right), \tag{12}$$

$$\frac{\mathrm{d}r_p}{\mathrm{d}t} = -\frac{D\,V_m}{\delta}\,\frac{r_p}{r_p + \delta}\left(C^{EQ}(T_1)\,\exp\left(\frac{2\sigma V_M}{R_G T_1 r_p}\right) - C_w\right). \tag{13}$$

We define the dimensionless variables

$$r_p = R_p\,\hat{r}_p\,, \qquad t = \tau\,\hat{t}\,, \qquad C_w = C_0\,\hat{C}_w\,, \tag{14}$$

where the unknown time scale τ is found presently. Substituting (14) into (12) yields

$$\frac{\mathrm{d}C_w}{\mathrm{d}t} = \tau\frac{3DR_p^2}{\delta r_w^3}\,r_p\,(r_p + \Delta)\left(H_C\,\exp\left(\frac{G_C}{r_p}\right) - C_w\right), \tag{15}$$

where the dimensionless notation has been dropped immediately, $\Delta = \delta/R_p = \mathcal{O}(1)$, $H_C \equiv \exp(B(T_1 - T_0)) = \mathcal{O}(1)$ and $G_C \equiv (2\sigma V_M)/(R_G R_p T_1)$. The last of these parameters is typically small for micron-sized particles. However, this is generally not the case for nanoparticles where R_p is small, and thus $G_C = \mathcal{O}(1)$. Hence, to keep the model general and applicable to particles of any size, we do not exploit the magnitude of G_C to make any reductions. We assume that the radius of the well-mixed region is related to the particle radius length scale via $r_w = \beta R_p$ where β is a measure of the distance between particles in the tank. The value of β depends on the diluteness of the solution where larger values of β yield larger well-mixed regions, and thus less lead sulphate nuclei in the tank. The time scale is found by balancing the parameters in (15) to give $\tau = (\beta^3 R_p \delta)/(3D)$. Thus,

$$\frac{\mathrm{d}C_w}{\mathrm{d}t} = r_p\,(r_p + \Delta)\left(H_C \exp\left(\frac{G_C}{r_p}\right) - C_w\right). \tag{16}$$

Introducing (14) into (13) leads to

$$\frac{\mathrm{d}r_p}{\mathrm{d}t} = -F_C\,\frac{r_p}{r_p + \Delta}\left(H_C \exp\left(\frac{G_C}{r_p}\right) - C_w\right), \tag{17}$$

where $F_C \equiv (\beta^3 V_M C_0)/3$. The initial conditions for each of the discussed temperature regimes are

$$C_w(0) = 1\,, \qquad r_p(0) = 1\,. \tag{18}$$

4.2. Time-dependent temperature models

Substituting the dimensionless time variable into the temperature functions (7) and (8) yields

$$T_L = \begin{cases} T_0 + (T_1 - T_0)t & \text{if } t \le 1 \\ T_1 & \text{if } t \ge 1\,, \end{cases} \tag{19}$$

$$T_E = T_1 - (T_1 - T_0)\,\mathrm{erfc}\,(t)\,, \tag{20}$$

where again the dimensionless notation has been negelcted. For $t \le 1$ the dimensionless linear temperature problem is

$$\frac{\mathrm{d}C_w}{\mathrm{d}t} = r_p\,(r_p + \Delta)\left(H_L(t) \exp\left(\frac{G_L}{T_L(t) r_p}\right) - C_w\right), \tag{21}$$

$$\frac{\mathrm{d}r_p}{\mathrm{d}t} = -F_C\,\frac{r_p}{r_p + \Delta}\left(H_L(t) \exp\left(\frac{G_L}{T_L(t) r_p}\right) - C_w\right), \tag{22}$$

where $G_L \equiv (2\sigma V_M)/(R_G R_p)$ and $H_L(t) \equiv \exp(B\,(T_1 - T_0)t)$. The initial conditions remain unchanged from the constant temperature case. For $t \ge 1$ the dimensionless linear problem is just the constant temperature equations (16) and (17) with $C_w(1) = C_{t_1}, r_p(1) = R_{t_1}$ where C_{t_1} and R_{t_1} are the lead concentration and the particle radius, respectively, at $t = 1$.

Similarly, the dimensionless error function temperature problem is

$$\frac{\mathrm{d}C_w}{\mathrm{d}t} = r_p\,(r_p + \Delta)\left(H_E(t) \exp\left(\frac{G_E}{T_E(t) r_p}\right) - C_w\right), \tag{23}$$

$$\frac{\mathrm{d}r_p}{\mathrm{d}t} = -F_C\,\frac{r_p}{r_p + \Delta}\left(H_E(t) \exp\left(\frac{G_E}{T_E(t) r_p}\right) - C_w\right), \tag{24}$$

where $G_E \equiv G_L$ and $H_E(t) \equiv \exp(B(T_1 - T_0)\mathrm{erfc}(t))$.

Table 1. Physical parameters used in simulations.

Meaning	Notation	Value	Units
Acceleration due to gravity	g	9.8	$\mathrm{m\,s^{-2}}$
Universal gas constant	R_G	8.31447	$\mathrm{J\,K^{-1}\,mol^{-1}}$
Lead sulphate density	ρ_{LS}	6290	$\mathrm{kg\,m^{-3}}$
Lead sulphate molar volume	V_M	4.821×10^{-5}	$\mathrm{m^3\,mol^{-1}}$
Lead diffusion coefficient	D	9.45×10^{-10}[13]	$\mathrm{m^2\,s^{-1}}$
Acid solution kinematic viscosity	ν_S	1.1×10^{-5}	$\mathrm{m^2\,s^{-1}}$
Acid solution dynamic viscosity	μ_S	0.021	$\mathrm{N\,s\,m^{-2}}$
Initial acid solution velocity	v_p	10	$\mathrm{m\,s^{-1}}$
Average lead sulphate particle radius	R_p	10^{-5}	m
Surface energy	σ	8.01×10^{-2} [14]	$\mathrm{J\,m^{-2}}$
Initial lead concentration	C_0	0.0128	$\mathrm{mol\,m^{-3}}$
Initial acid solution temperature	T_0	333	K
Final acid solution temperature	T_1	283	K
Equilibrium concentration fitting parameters	A, B	$2 \times 10^{-6}, 0.03$	$\mathrm{mol\,m^{-3}}$, $\mathrm{K^{-1}}$
Width of diffusion region	δ	10^{-4}	m
Well mixed region parameter	β	75	-

5. Results

In this section the results of the model are discussed. The equations for the fixed and time-dependent temperature cases were simulated numerically in Matalab. Table 1 summarises the main physical and dimensionless parameters of the problem. The temperature-dependent equilibrium concentration $C^{EQ}(T) = A\exp(BT)$ is obtained by fitting the function with the experimental data of Crockford and Brawley [11, 12] for lead sulphate solubility over a range of temperatures and sulphuric acid concentrations in aqueous solutions. The width of the diffusion region around the particle is approximated via Prandtl's boundary layer (see Appendix A). The well mixed-region parameter $\beta = \rho_w/R_p$ is obtained by considering the distance between particles (i.e., solution diluteness) and the average particle radius. Based on experimental data from our industrial collaborator we found $\beta = 75$.

Figure 2 shows numerical solutions for the dimensionless concentration and particle radius subject to the fixed temperature (blue), linear temperature profile (green) and error function temperature configurations. We note that the difference between the three temperature configurations with respect to growth is negligible. The graphs demonstrate the correct qualitative behaviour where the lead sulphate particle grows via the flux of lead atoms, and hence the lead concentration in the well-mixed region decreases. Particle growth ends when a dynamic equilibrium between the particle and the lead atoms in the acid solution is attained. The simulations show that after 3 units of dimensionless time (or approximately 11.5 h) the concentration in the well-mixed region reaches a steady state and the particle has achieved its maximum size. Figure 2(b) also shows that there is a minimal increase of approximately 1.7% in particle radius during growth. This increase can be attributed to the small dimensionless parameter F_C in the particle radius equation. The F_C parameter depends on β where we note that larger values of β indicate the presence of less lead sulphate particles and consequently, less competition for lead atoms, which in turn leads to larger lead sulphate particles. However, the cubic dependence of the time scale τ on β implies that larger nuclei require much longer to grow.

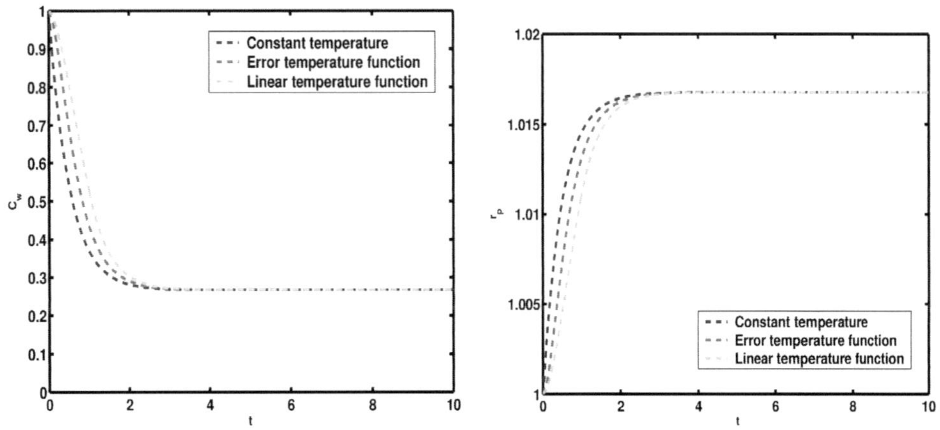

(a) (b)

Figure 2. Evolution of (a) well-mixed region lead concentration and (b) lead sulphate particle radius subject to constant temperature (blue), error function temperature (red) and linear temperature function(green).

The critical radius evolution for each of the temperature configurations is shown in Figure 3. The curves divide regions which support particle growth (above curves) and particle dissolution (below curves). Initially the time-dependent temperature configurations have higher critical radii than that of the constant temperature regime. Thus, the former temperature set-ups are preferable as it is more difficult for new particles to precipitate out of solution due to the higher critical energy barrier. Unlike the other two curves, the constant temperature curve is monotonically increasing. For the constant temperature case when the acid solution enters the tanks its temperature decreases rapidly. This causes the lead sulphate solubility to decrease dramatically and thus the initial particle critical radius at $t = 0$ drops instantaneously to its minimum value. On the other hand, for the time-dependent temperature cases, the acid solution is subject to a slower cooling regime. Hence, the lead sulphate solubility limit decreases at a slower rate and reaches it minimum after approximately 0.7 units of dimensionless time (or 2.7 h). As the settling process proceeds to the point and beyond where the constant temperature curve intersects with the other two curves, we can see that the constant temperature configuration is an improvement on the other two temperature regimes.

6. Conclusion

In this paper we investigated the effect that temperature has on a lead sulphate settling process from the glassware manufacturing industry. A similar strategy to that proposed by Liger-Belair *et al.* [9] for modelling bubble nucleation from cellulose fibers in carbonated beverages was applied. The spherical region around the particle was divided into two separate regions each with its own characteristic lead atom diffusion regime. Specifically, the relatively small boundary layer region in contact with the particle was characterised by position-dependent diffusion and the larger (fast diffusion), far field well-mixed region was typified by time-dependent lead diffusion (slow diffusion). The final model consisted of two coupled nonlinear ordinary differential equations for the particle radius and the far-field concentration.

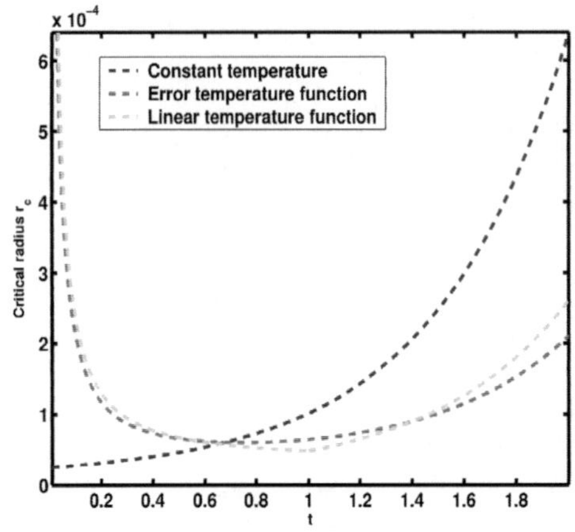

Figure 3. Evolution of dimensionless critical radius subject to constant temperature (blue), error function temperature (red) and linear temperature function (green).

Three temperature configurations were considered. The numerical simulations demonstrated minor particle growth. Furthermore, the time required for the particle to reach its maximum size was of the order hours. From an operating perspective, the acid solution should be allowed to sit overnight to maximise lead sulphate extraction. The small lead diffusion coefficient is one reason for the length in time required to achieve maximum particle growth. The introduction of a stirring mechanism, at least of the upper solution layers, could speed up the lead diffusion process.

An expression for the particle critical radius was also formulated. The time-dependent temperature configurations were shown to improve on the constant temperature case near the start of the process where particles had more difficulty overcoming the critical energy barrier associated with the former temperature regimes compared to the latter. However, as settling continues the performance of the constant temperature configuration was shown to be the best of the three configurations with respect to the dissolution of smaller particles and the growth of existing crystals.

Acknowledgments
We gratefully acknowledge the financial support of the Mathematics Applications Consortium for Science and Industry (MACSI, http://www.macsi.ul.ie) supported by a Science Foundation Ireland mathematics initiative Grant 06/MI/005 and an Embark Initiative postgraduate Award RS/2006/41.

Appendix A. Boundary layer width

Appendix A.1. Introduction

In this section we quantify the width of the diffusion region, δ around the lead sulphate particle. In Appendix A.2 we apply Prandtl's boundary theory to obtain an approximate expression for δ. Prandtl's approximation depends on the velocities of the acid solution and the particle, both of which are random in nature. Thus, in Appendix A.3 we formulate a set of stochastic differential equations for their velocities. In Appendix A.4 we derive a Fokker-Planck equation for the joint probability density function for the root mean square particle-solution relative velocity, which is then used to find δ.

Appendix A.2. Boundary layer approximation
We approximate the width of the particle diffusion region via Prandtl's boundary layer theory [15], which states $\delta \sim \sqrt{(\nu L)/U}$ where ν is the liquid kinematic viscosity, L is a characteristic length scale associated with the body which the liquid is flowing past and U is a liquid velocity scale. Thus, for the current study

$$\delta \sim \sqrt{\frac{\nu_S R_p}{u - v}}\,, \tag{A.1}$$

where ν_S is the acid solution dynamic viscosity and u, v are the velocities of the acid solution and particle, respectively. Assuming that the particle has a velocity component only in the vertical direction and negligible buoyancy, then via Newton's second law of motion, the one-dimensional equation for motion of a single spherical particle is

$$M_p \frac{\mathrm{d}v}{\mathrm{d}t} = -M_p\, g - F_D\,, \tag{A.2}$$

where g is gravity and M_p is particle mass. The drag force, F_D, depends on the acid solution flow regime, which in turn itself is charecterised by the magnitude of the Reynolds number $\mathrm{Re} = ((U - V)R_p)/\nu_S$. Using the parameter values in Table 1 and assuming that the difference between the acid solution and particle velocity scales U, V is $O(1)$ or less, then Re is presumed to be sufficiently small such that Stokes' law for the drag force is applicable. Thus, from [16]

$$F_D = -6\,\pi\,\mu_S\,R_p\,(v - u)\,, \tag{A.3}$$

where μ_S is the solution kinematic viscosity. Upon substituting (A.3) into (A.2) we obtain

$$\frac{\mathrm{d}v}{\mathrm{d}t} = -g - \frac{1}{\tau_1}(v - u)\,, \tag{A.4}$$

where $\tau_1 = M_p/(6\pi\mu_S R_p) \ll 1\,\mathrm{s}$ is a measure of the response time of the particle to interactions with the acid solution.

Appendix A.3. Acid solution velocity
The random motion of the acid velocity, and thus the interactions of the acid solution with the particle, is modelled via a time-dependent stochastic function. Specifically, we assume the solution velocity can be represented by a Gaussian stochastic function $u(t)$ with zero mean, variance $\langle u^2 \rangle = v_p^2$ and autocorrelation function

$$R(\delta t) = \langle\, u(t)\, u(t + \delta t)\,\rangle = v_p^2 \exp\left(-\delta t/\tau_2\right), \tag{A.5}$$

where we presume that the root mean squared acid solution velocity is equal to the initial velocity of the acid solution v_p (as imparted by the pump), $\delta t\ (> 0)$ is the time lag and τ_2 is the correlation time [17]. For the current system $\tau_2 = d/v_p$ where $v_p \approx 10\,\mathrm{m\,s^{-1}}$ is the acid solution velocity due to the pump and $d \equiv \sqrt{\nu_S\, t_{\mathrm{res}}} \approx 0.12\,\mathrm{m}$ is a length scale related to the size of the turbulent eddies in the solution. The time $t_{\mathrm{res}} \approx 1\,\mathrm{hr}$ corresponds to the residence time of the acid solution in the settlement tank and thus $\tau_2 \approx 0.03\,\mathrm{s}$.

To derive the stochastic differential equation (SDE) for u we begin by defining the realisation of the acid solution velocity at time $t + \delta t$ in terms of a deterministic and a completely random component. The deterministic part is dependent on the realisation of the acid solution at t and the time lag δt. Thus,

$$u(t + \delta t) = g(\delta t)\, u(t) + \sigma\, \xi(t)\,, \tag{A.6}$$

where $g(\delta t)$ is an unknown function and σ is an unknown constant known as the average velocity per square-root time of the random fluctuations. The $\xi(t)$ term in (A.6) represents a completely random white noise process with zero mean. The autocorrelation function associated with $\xi(t)$ is $R(t) = \alpha^2\delta(t)$ where α^2 is some positive constant and $\delta(t)$ is the standard Dirac delta function. We note that $\xi(t)$ is the time derivative of a Wiener process $W(t)$ such that

$$\xi(t) = \frac{\mathrm{d}W}{\mathrm{d}t}. \tag{A.7}$$

An expression for $g(\delta t)$ is obtained by multiplying $u(t)$ by $u(t+\delta t)$ and averaging over the ensemble of realisations to give

$$\langle\, u(t)\, u(t+\delta t)\,\rangle = \langle\, u(t)\,(g\, u(t) + \sigma\xi(t))\,\rangle = g\, v_p^2\,. \tag{A.8}$$

Comparing (A.5) with (A.8) it follows that $g(\delta t) = 1 - \delta t/\tau_2 + O(\delta t^2)$ for $t \ll 1$. Expanding (A.6) for small δt, diving through by δt and multiplying by $\mathrm{d}t$ yields

$$\mathrm{d}u = -\frac{u}{\tau_2}\,\mathrm{d}t + \sigma'\,\mathrm{d}W\,, \tag{A.9}$$

where $\sigma' = \sigma/\delta t$ and (A.7) has been exploited. To find σ' we first obtain $\langle u^2\rangle$ via Itô's lemma [17], which states that for a SDE of the form

$$\mathrm{d}x = a(x,t)\,\mathrm{d}t + b(x,t)\,\mathrm{d}W\,, \tag{A.10}$$

and any twice differentiable function $f(x,t)$ of two real variables x and t, then

$$\mathrm{d}f(x,t) = \left(\frac{\partial f}{\partial t} + a(x,t)\frac{\partial f}{\partial x} + \frac{1}{2}\,b(x,t)^2\frac{\partial^2 f}{\partial x^2}\right)\mathrm{d}t + b(x,t)\frac{\partial f}{\partial x}\,\mathrm{d}W\,. \tag{A.11}$$

Hence, defining the new variable $f(u) = u^2$ and applying Itô's lemma to (A.9) yields

$$\mathrm{d}(u^2) = \left(-\frac{2u^2}{\tau_2} + \sigma'^{\,2}\right)\mathrm{d}t + 2\,u\,\sigma'\,\mathrm{d}W\,. \tag{A.12}$$

An expression for σ' is found by considering a general mean-reverting stochastic process [18]

$$\mathrm{d}S = a\,(L - S)\,\mathrm{d}t + \sigma\,\mathrm{d}W\,, \tag{A.13}$$

where a is the rate of mean-reverting. The solution to (A.13) tends to the equilibrium value, L with the stochastic term causing S to oscillate around L. Hence, (A.12) can be viewed as an equation for a mean-reverting stochastic process with $S = -(2u^2)/\tau_2$ and $L = \sigma'^{\,2}$, respectively. Thus, comparing terms in (A.12) yields $\sigma'^2 = (2\,u^2)/\tau_2$ or equivalently

$$\sigma' = v_p\sqrt{\frac{2}{\tau_2}}\,, \tag{A.14}$$

where we have exploited the fact that the magnitude of u^2 is $O(v_p^2)$. Introducing (A.14) into (A.9) yields the SDE

$$\mathrm{d}u = -\frac{u}{\tau_2}\,\mathrm{d}t + v_p\sqrt{\frac{2}{\tau_2}}\,\mathrm{d}W\,. \tag{A.15}$$

We define the dimensionless variables

$$v = v_p\,\hat{v} - \tau_1 g\,, \quad u = v_p\hat{u}\,, \quad t = \tau_2\hat{t}\,, \quad W = \sqrt{\frac{\tau_2}{2}}\hat{W}\,, \tag{A.16}$$

which upon substitution into (A.4) and (A.15) yields

$$\mathrm{d}v = \alpha\,(u - v)\,\mathrm{d}t\,, \qquad \mathrm{d}u = -u\,\mathrm{d}t + \mathrm{d}W\,, \tag{A.17}$$

where the dimensionless notation has been dropped and $\lambda = \tau_2/\tau_1$.

Appendix A.4. Fokker-Planck equation and solution

We now formulate the Fokker-Planck associated with the joint probability density function of u and v. An arbitrary system of two SDEs in $x_1(t)$ and $x_2(t)$ is given by [19]

$$\frac{\mathrm{d}x_1}{\mathrm{d}t} = f_1(x_1, x_2) + g_1(x_1, x_2)\eta_1(t)\,, \qquad \frac{\mathrm{d}x_2}{\mathrm{d}t} = f_2(x_1, x_2) + g_2(x_1, x_2)\eta_2(t)\,, \tag{A.18}$$

where $x_2(t)$ has autocorrelation $< x_2(t)x_2(t') >= \kappa \exp(|tt'|/\tau)$ and $\eta_1(t), \eta_2(t)$ are random force terms. The related two-dimensional Fokker-Planck equation for the joint probability density function of $x_1(t)$ and $x_2(t)$, P, is

$$\frac{\partial P}{\partial t} = -\frac{\partial}{\partial x_1}\Big(f_1\,P\Big) - \frac{\partial}{\partial x_2}\Big(f_2 P\Big) + \kappa\frac{\partial}{\partial x_1}\Big(g_1\frac{\partial}{\partial x_1}(g_1 P)\Big) + \kappa\,\frac{\partial}{\partial x_2}\Big(g_2\frac{\partial}{\partial x_2}\,(g_2 P)\Big). \tag{A.19}$$

Hence, using $x_1 = v$, $x_2 = u$, $f_1 = \lambda(u - v)$, $f_2 = -u$, $g_1 = 0$ and $g_2 = 1$, the Fokker-Planck equation associated with (A.17) is

$$\frac{\partial}{\partial u}\Big(uf\Big) + \lambda\frac{\partial}{\partial v}\Big((v-u)f\Big) + \frac{\partial^2 f}{\partial u^2} = 0\,. \tag{A.20}$$

where f is the stationary joint probability density function of u and v. If the first term in (A.20) is neglected, the resulting reduced equation has as its solution $g(u) = \exp(-u^2/2)$. Thus, we assume that the full solution to (A.20) can be written as $f(u, v) = g(u)h(u, v)$. The function $h(u, v)$ is determined via heuristic physical considerations regarding the nature of $f(u, v)$, and in particular the effect of the parameter λ on the relationship between u and v in (A.17). Firstly, inspection of the second SDE in (A.17) suggests that if $\lambda \gg 0$, then the particle velocity v tends towards the acid solution velocity u. Thus, in the limit as $\lambda \to \infty$ there is perfect correlation between u and v. On the other hand, as $\lambda \to 0$, the autocorrelation between u and v vanishes and v oscillates around its mean value. Assuming that the function h is exponential in nature, for intermediary values of λ we have $h(u, v) = \exp(-B[v - Au]^2)$ where A and B are unknown constants. Based on the above definition of $f(u, v)$, the function takes the form $f(u, v) = \exp(-u^2/2 - B[v - Au]^2)$. The constants A and B are found by substituting this expression for f into the Fokker-Planck equation and matching like coefficients for u and v to obtain

$$f(u, v) = \exp\Big(-\frac{u^2}{2} - \frac{\lambda}{2}\Big[u - \frac{\lambda+1}{\lambda}v\Big]^2\Big)\,. \tag{A.21}$$

Finally, we average over the joint probability distribution of u and v to obtain the mean square particle relative velocity

$$\langle (u-v)^2\rangle = \frac{\int_{-\infty}^{\infty}\int_{-\infty}^{\infty}(v-u)^2 f(u, v)\,\mathrm{d}u\,\mathrm{d}v}{\int_{-\infty}^{\infty}\int_{-\infty}^{\infty} f(u, v)\,\mathrm{d}u\,\mathrm{d}v} = \frac{1}{\lambda+1}\,, \tag{A.22}$$

which can be rewritten in dimensional terms via (A.16) and then used in (A.1).

References

[1] Montgomery JM 1985 *Water Treatment: Principles and Design* (John Wiley and Sons).
[2] Galvin KP 2010 Gravity separation of coal in the reflux classifier - new mechanisms for suppressing the effects of particle size *Int. J. Coal Prep. Util.* **30** 130-44.
[3] Laplante AR, Woodcock F and Noaparast M 1995 Predicting gravity separation gold recoveries *Miner. Metall. Proc.* **12** 74-9.

[4] Vettore L, De Matteis MC and Zampini P 1980 A new density gradient system for the separation of human red blood cells *Am. J. Hemat.* **8** 291-7.
[5] Ward J, Fowler AC and O'Brien SBG 2011 Acid polishing of lead glass *J. Math. Ind.*, **1** 1.
[6] Li Y, Kang JH, Lau SL, Kayhanian M and Stenstrom MK 2007 Optimization of settling tank design to remove particles and metals *J. Environ. Eng.-ASCE* **134** 885-94.
[7] Boon AG and Dolan JF 1995 Design of settlement tanks and the use of chemicals to aid precipitation of suspended solids *J. Chart. Inst. Water E* **9** 57-68.
[8] Agridiotis V, Forster CF and Carliell-Marquet C 2007 Addition of Al and Fe salts during treatment of paper mill effluents to improve activated sludge settlement characteristics *Bioresource Technol.* **98** 2926-34.
[9] Liger-Belair G, Voison C and Jeandet P 2005 Modeling nonclassical heterogeneous bubble nucleation from cellulose fibers: application to bubbling in carbonated beverages *J. Phys. Chem.* **109** 14574-80.
[10] Park J, Joo J, Kwon SG, Jang Y and Hyeon T 2007 Synthesis of monodisperse spherical nanocrystals *Angew. Chem-Ger. Edit.* **46** 4630-60.
[11] Crockford HD and Brawley DJ 1934 The solubility of lead sulfate in water and aqueous solutions of sulphuric acid *J. A. Chem. Soc.* **56** 2600-1.
[12] Crockford HD and Brawley DJ 1936 The solubility of lead Sulfate in aqueous solutions of sulfuric acid at high concentrations *J. Phys. Chem.-US* **40** 303-5.
[13] Lide DR 2006 *RC Handbook of Chemistry and Physics: A Ready Reference Book of Chemical and Physical Data* (Boca Raton, FL. , London).
[14] Bernardi DM 1990 Nucleation of lead sulfate in porous lead-dioxide electrodes *J. Electrochem. Soc.* **137** 1670-81.
[15] Ackroyd JAK, Axcell BP and Ruban AI 2001 *Early Developments in Modern Aerodynamics* (Butterworth-Heinemann Oxford).
[16] Fowler AC 1997 *Mathematical Models in the Applied Sciences* (Cambridge Texts in Applied Mathematics).
[17] Gardiner CW 2003 *Handbook of Stochastic Methods* (Springer Berlin).
[18] Van Kampen NG 2007 *Stochastic Processes in Physics and Chemistry* Elsevier.
[19] Risken H 1996 *The Fokker-Planck Equation: Methods of Solutions and Applications* (Springer, Berlin).

Conditions for the critical phenomena in a dynamic model of an electrocatalytic reaction

Natalia Firstova, Elena Shchepakina
Department of Technical Cybernetics, Samara National Research University, 34, Moskovskoye shosse, Samara 443086, Russian Federation
E-mail: firstova.natalia@yandex.ru, shchepakina@smr.ru

Abstract. In the paper, a detailed study of the critical phenomena mechanism in a model of an electrocatalytic reaction is carried out. The use of the geometric theory of integral manifolds allows us to show the relationship between the critical regime and the stability loss delay of the slow integral manifold of the corresponding differential system. The asymptotic formula for the critical value of the control parameter is obtained.

1. Introduction

In this paper we investigate the Koper–Sluyters electrocatalytic reaction mechanism underlying an electrochemical reactor. The electrochemical reactor, in contrast to its closest analogue — diaphragm electrolyzer, has much more technical and technological degrees of freedom and has been designed for the electrochemical transformation of a variety of liquids (not only water or aqua-systems over a wide range of solute concentrations, but also milk, vegetable and mineral oils, solutions of carbohydrates, ammonia, alcohols, surfactants, organic and inorganic fertilizers, herbicides, pesticides, and many others). The application area of electrochemically activated solutions is quite broad and includes medicine, petrochemical industry, manufacture of beverages, drinking water purification.

The Koper–Sluyters electrocatalytic reaction is a chemical reaction corresponding to the following kinetic scheme [1]:

$$X_{bulk} \xrightarrow{D/\delta} X_{sur} \underset{k_d}{\overset{k_a}{\rightleftharpoons}} X_{ads} \xrightarrow{k_e} P + ne^- .$$

Here X is a single species which diffuses towards the electrode, where it successively adsorbs and is electrochemically oxidized; X_{sur} and X_{bulk} are the parts of X on the electrode surface and outside the electric double layer, respectively; X_{ads} is the adsorbed part of X; D is the diffusion coefficient of X; δ is the thickness of the Nernst diffusion layer; k_a, k_d, k_e are the rate constants for adsorption, desorption and electron transfer, respectively. The oxidation products P are assumed not to be adsorbed nor to leave neighborhood of the interface.

In paper [1], simple models of the Koper-Sluyters reaction were investigated in order to determine the possible origins of instabilities and oscillations. For proposed models three different kinds of external electrochemical cell control were considered: potentiostatic control, ohmic resistance control and galvanostatic control. The relations of the proposed models to the previously studied models and experiments were also discussed here.

The case of potentiostatic control of the Koper–Sluyters reaction was also investigated in [2]. The authors showed the existence of two types of the oscillations, small amplitude harmonic oscillations and large amplitude relaxation oscillations. It was noted that for a very narrow range of electrode potentials the electrocatalytic reaction behaviour is similar to a canard. However, neither numerically nor analytically the canard's existence has been shown. The aim of the present paper is to fill this gap.

The dimensionless dynamic model of the Koper–Sluyters reaction in the case of potentiostatic control is described by the system

$$\frac{du}{dt} = -k_a e^{\gamma\theta/2} u(1-\theta) + k_d e^{-\gamma\theta/2}\theta + 1 - u = f(u,\theta), \tag{1}$$

$$\beta\frac{d\theta}{dt} = k_a e^{\gamma\theta/2} u(1-\theta) - k_d e^{-\gamma\theta/2}\theta - k_e e^{\alpha_0\zeta E}\theta = g(u,\theta), \tag{2}$$

here u is the dimensionless interfacial concentration of X; θ is the dimensionless amount of X that is adsorbed on the electrode surface; E is the interfacial potential; β is the coverage ratio of the adsorbate; α_0 is the symmetry factor for the electron transfer; and $\zeta = F/(RT)$, where R is the universal gas constant, F is Faradays constant, and T is the temperature. The physical meaning of the parameter γ has always been a subject of dispute. In most of the literature it is interpreted as an interaction parameter. Positive γ signifies attractive and negative γ signifies repulsive adsorbate interactions.

The typical values of the parameter β are small [1,2], hence, the geometric theory of singular perturbations [3–5] can be applied for the qualitative study of (1), (2).

2. Analysis of the slow integral manifold

Recall, that a slow integral manifold of a singular perturbed system is an invariant surface of slow motions whose dimension is equal to that of the slow subsystem [4, 5].

The zero-order approximation of the slow integral manifold, the *slow curve*, of the system (1), (2) is determined by the expression $g(u,\theta) = 0$ and has the form

$$k_a e^{\gamma\theta/2} u(1-\theta) - k_d e^{-\gamma\theta/2}\theta - k_e e^{\alpha_0\zeta E}\theta = 0$$

or

$$u = \frac{(k_d e^{-\gamma\theta/2} + k_e e^{\alpha_0\zeta E})\theta}{k_a e^{\gamma\theta/2}(1-\theta)}. \tag{3}$$

The slow curve has a stable (or attractive) and unstable (or repulsive) parts, which are zero-order approximations of the stable and unstable slow integral manifolds, respectively. The stable and unstable parts are divided by jump points [4, 5], at which we have

$$\begin{cases} g(u,\theta) = 0, \\ \dfrac{\partial g(u,\theta)}{\partial\theta} = 0. \end{cases} \tag{4}$$

For (1), (2) the system (4) has the form

$$\begin{cases} k_a e^{\gamma\theta/2} u(1-\theta) - k_d e^{-\gamma\theta/2}\theta - k_e e^{\alpha_0\zeta E}\theta = 0, \\ k_a u(1-\theta)\frac{\gamma}{2} e^{\gamma\theta/2} - k_a u e^{\gamma\theta/2} - k_d e^{-\gamma\theta/2} + k_d e^{-\gamma\theta/2}\theta\frac{\gamma}{2} - k_e e^{\alpha_0\zeta E} = 0. \end{cases} \tag{5}$$

The shape of the slow curve depends on the values of the parameters. We will consider γ as a control under the fixed values of all other parameters.

The following cases are possible.

For $0 < \gamma < 4$ the system (5) has no solutions. Therefore, the slow manifold is either entirely stable or entirely unstable. Due to the fact that

$$\frac{\partial g(u,\theta)}{\partial \theta} < 0,$$

i.e.,

$$k_a u(1-\theta)\frac{\gamma}{2}e^{\gamma\theta/2} - k_a u e^{\gamma\theta/2} - k_d e^{-\gamma\theta/2} + k_d e^{-\gamma\theta/2}\theta\frac{\gamma}{2} - k_e e^{\alpha_0 \zeta E} < 0,$$

the slow curve is stable in this case, see Figure 1. Hence, the trajectories of the system (1), (2) are attracted to the slow curve and then follow along it as $t \to \infty$. In the case when $\gamma \approx 4$, the slow curve is stable too [6] and has a shape represented in Figure 2.

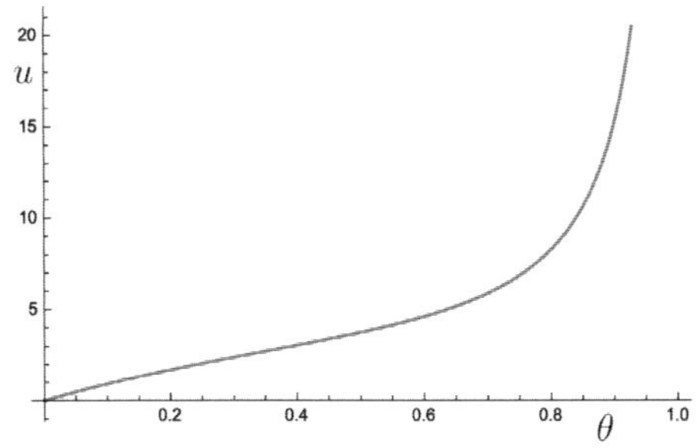

Figure 1. The slow curve of the system (1), (2) for $\gamma = 2$.

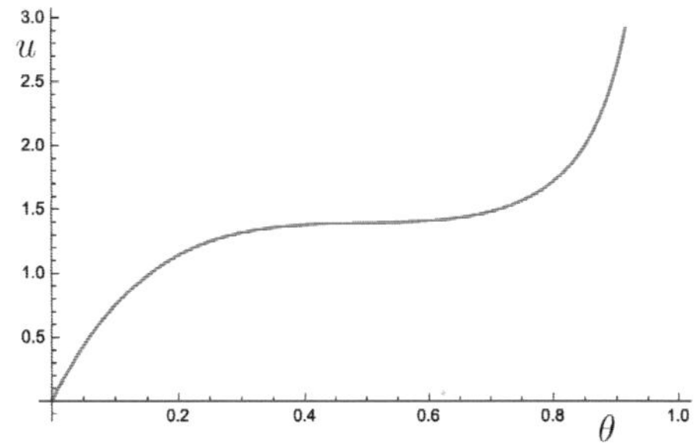

Figure 2. The slow curve of the system (1), (2) for $\gamma = 4$.

Thus, for $\gamma > 4$ two jump points, A_1 and A_2, divide the slow curve into three parts S_1^s, S_2^u, S_3^s (see Figure 3) which are zeroth order approximations for the corresponding slow integral manifolds M_1^s, M_2^u, M_3^s. Manifolds M_1^s and M_3^s are stable and M_2^u is unstable. Each manifold M_1^s, M_2^u, and M_3^s is at the same time part of some trajectory of the system (1), (2) [5].

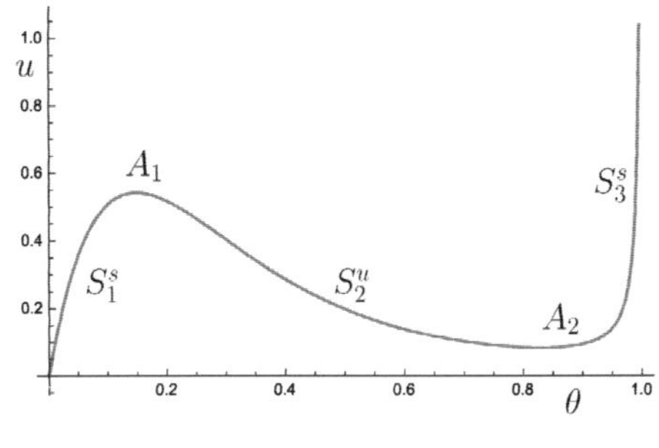

Figure 3. The slow curve of the system (1), (2) for $\gamma = 8$. The parts S_1^s and S_3^s are stable while S_2^u is unstable, A_1 and A_2 are the jump points.

A trajectory, starting from an initial point in the basin of attraction of the stable slow integral manifold M_1^s (or M_3^s), tends rapidly to it and then follows M_1^s (or M_3^s) with the velocity of order $O(1)$ as $\beta \to 0$. The further behaviour of the trajectory will depend on the location of the critical point of the system (1), (2).

The critical points of the system (1), (2) are determined by the system

$$\begin{cases} -k_a e^{\gamma\theta/2}u(1-\theta) + k_d e^{-\gamma\theta/2}\theta + 1 - u = 0, \\ k_a e^{\gamma\theta/2}u(1-\theta) - k_d e^{-\gamma\theta/2}\theta - k_e e^{\alpha_0\zeta E}\theta = 0. \end{cases} \tag{6}$$

From the system (6) we get:

$$u = 1 - k_e e^{\alpha_o \zeta E}\theta. \tag{7}$$

Substituting (7) into the second equation of the system (6), we obtain the equation that determines the value $\theta = \theta^*$:

$$k_a e^{\gamma\theta/2}(1 - k_e e^{\alpha_o\zeta E}\theta)(1-\theta) - k_d e^{-\gamma\theta/2}\theta - k_e e^{\alpha_o\zeta E}\theta = 0. \tag{8}$$

Thus, we obtain the critical point

$$A\left(\theta^*, 1 - k_e e^{\alpha_o\zeta E}\theta^*\right),$$

where θ^* is the solution of the equation (8).

The Jacobian matrix of the system (1), (2):

$$\begin{pmatrix} \frac{\partial f(u,\theta)}{\partial u} & \frac{\partial f(u,\theta)}{\partial \theta} \\ \frac{\partial g(u,\theta)}{\partial u} & \frac{\partial g(u,\theta)}{\partial \theta} \end{pmatrix},$$

has the characteristic equation

$$\lambda^2 + \lambda\xi_1 + \xi_2 = 0,$$

with the discriminant

$$D = \xi_1^2 - 4\xi_2,$$

where

$$\xi_1 = \frac{k_a}{\beta}e^{\gamma\theta^*/2}\left(1 - k_e e^{\alpha_0\zeta E}\theta^*\right)\left(1 - \frac{\gamma}{2}(1-\theta^*)\right) + \frac{k_d}{\beta}e^{-\gamma\theta^*/2}\left(1 - \frac{\gamma\theta^*}{2}\right)$$
$$+\frac{k_e}{\beta}e^{\alpha_0\zeta E} + k_a e^{\gamma\theta^*/2}(1-\theta^*) + 1,$$

$$\xi_2 = \frac{k_a}{\beta}e^{\gamma\theta^*/2}\left(1 - k_e e^{\alpha_0\zeta E}\theta^*\right)\left(1 - \frac{\gamma}{2}(1-\theta^*)\right) + \frac{k_d}{\beta}e^{-\gamma\theta^*/2}\left(1 - \frac{\gamma\theta^*}{2}\right)$$
$$+\frac{k_e}{\beta}e^{\alpha_0\zeta E} + \frac{k_a k_e}{\beta}e^{\gamma\theta^*/2}(1-\theta^*)e^{\alpha_0\zeta E}.$$

The type of the critical point and its coordinates depends on the value of the parameter γ. Let us consider the most interesting case when $\gamma > 4$. Without loss of generality the parameters of the system are chosen to be $\epsilon = 0.2$, $\gamma = 8.99$, $k_a = 10$, $k_d = 100$, $\alpha_0 = 0.05$, $f = 38,7$, $E = 0.207564$ unless other values are specified in figure captions.

In [6], it has been shown that the critical point is a stable focus when it lies on the stable part of the slow curve (Figure 4) and it is an unstable focus when it situated on S_2^u. In the second case, the relaxation oscillations are observed in the system, see Figure 5.

The transition between these two situations corresponds to the case when the critical point coincides with the jump point, the stable equilibrium of the system becomes unstable, and at the same instant the stable limit cycle is originated, i.e., the Andronov–Hopf bifurcation occurs, see Figure 6. With further minor modifications of the control parameter, say k_e, the critical point shifts on the unstable part of the slow curve, staying in small (of order $O(\beta)$ as $\beta \to 0$) neighbourhood of the jump point. As parameter k_e changes further, this limit cycle grows, and at a value $k_e = k_e^*$ (so-called canard point) it becomes the canard cycle [7] (Figures 7, 8) with the following canard explosion [8–10]. Recall that the trajectories which at first move along the stable slow integral manifold and then continue for a while along the unstable slow integral manifold are called canards [4, 5].

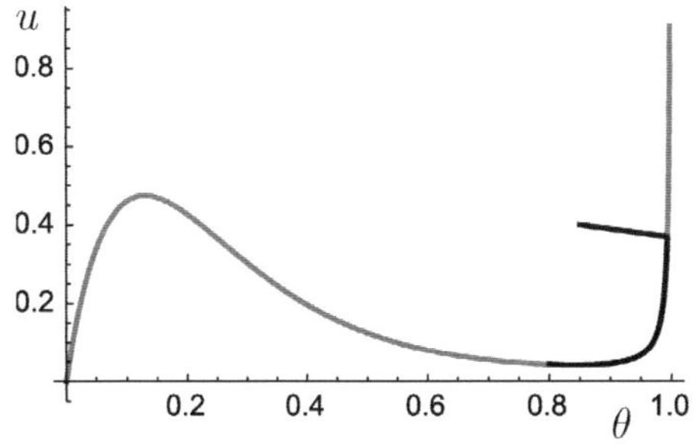

Figure 4. The slow curve (the red line) and the trajectory (the black line) of the system (1), (2); $k_e = 0.8$.

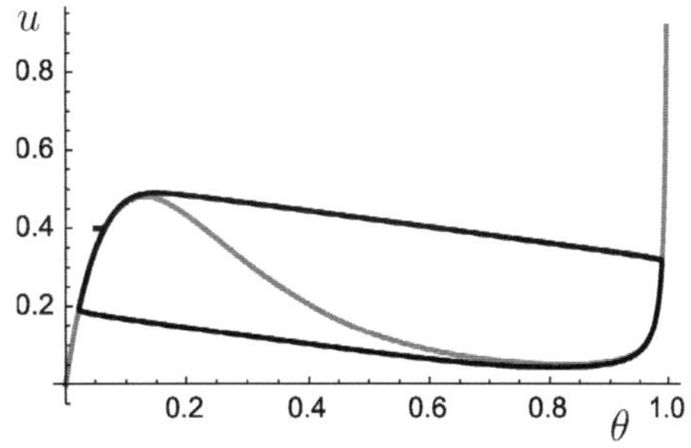

Figure 5. Relaxation oscillations in the system (1), (2); $k_e = 1.285$.

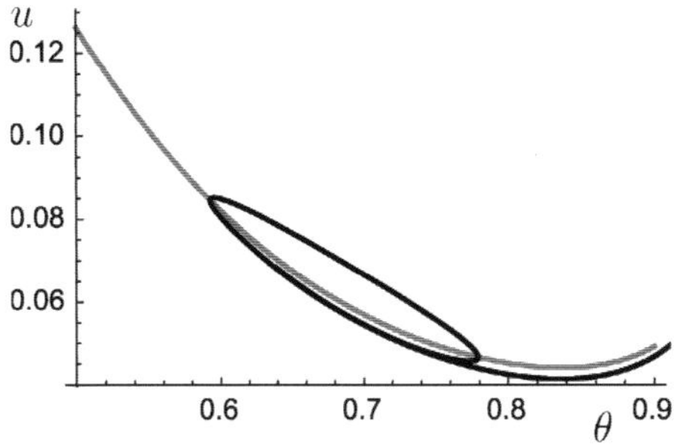

Figure 6. The Andronov-Hopf bifurcation in the system (1), (2); $k_e = 0.92$.

At the first sight the threshold in the qualitative behaviour of the solutions of the system corresponds to the Andronov–Hopf bifurcation point. However, when the value of the control parameter is close to the Andronov–Hopf bifurcation point, the size of the limit cycle is so small that the behaviour of the system' solution is practically indistinguishable from the slow mode. If, in the case of slow regime, the trajectories approach the stable equilibrium, practically coinciding with the jump point, in the later case they tend to a small limit cycle, again nearly coinciding with the same jump point. Only when the control parameter attains the canard point, provided the equilibrium is on the unstable part of the slow curve and in the sufficiently small vicinity of the jump point, the qualitative change of the system's behaviour can be observed. Namely, the growth of the limit cycle occurs in such a way that it becomes possible to speak of the existence of the canard trajectory. In other words, the appreciable change in size and/or in form of the limit cycle is observed for small variations of the control parameter, i.e. the canard explosion takes place. Thus, the canard point is the critical value of the control parameter.

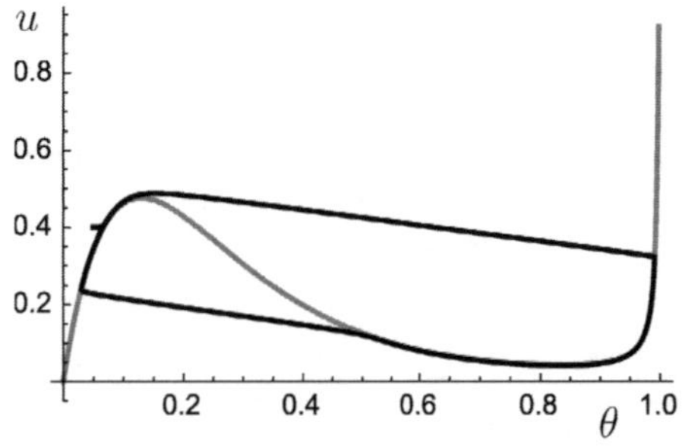

Figure 7. The slow curve (the red line) and the canard (the black line) of the system (1), (2); $k_e = 0.9205283$.

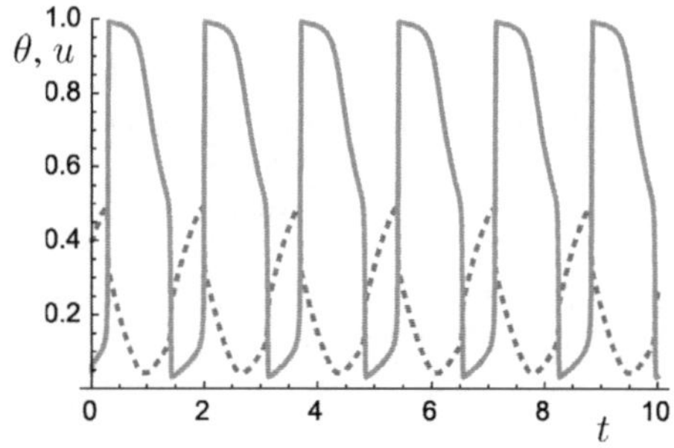

Figure 8. Plots of $\theta(t)$ (the solid orange line), $u(t)$ (the dashed blue line) for $k_e = 0.9205283$.

3. Asymptotic expression for the critical value of the control parameter

The canards and the parameter value k_e^* allow asymptotic expansions in powers of the small parameter β [4, 5, 11]:

$$u = \Phi(\theta, \beta) = u_0(\theta) + \beta u_1(\theta) + \beta^2 u_2(\theta) + \dots, \tag{9}$$

$$k_e^* = \chi(\beta) = \chi_0 + \beta\chi_1 + \beta^2\chi_2 + \dots . \tag{10}$$

In order to find these asymptotic expansions for the canard and the canard point we substitute the formal expansions (9) and (10) into the invariance equation [5]

$$\frac{du}{d\theta} g(u, \theta) = \beta f(u, \theta).$$

which follows from the system (1), (2). As a result we obtain the following equation:

$$\left(k_a e^{\gamma\theta/2}(1-\theta)(u_0 + \beta u_1 + \beta^2 u_2 + \dots) - (\chi_0 + \beta\chi_1 + \beta^2\chi_2 + \dots)e^{\alpha_0\zeta E}\theta - k_d e^{\gamma\theta/2}\theta\right)$$
$$\times(u_0' + \beta u_1' + \beta^2 u_2' + \dots) = -\beta k_a e^{\gamma\theta/2}(1-\theta)(u_0 + \beta u_1 + \beta^2 u_2 + \dots)$$
$$+\beta\left(k_d e^{\gamma\theta/2}\theta + 1 - u_0 - \beta u_1 - \beta^2 u_2 + \dots\right). \tag{11}$$

On setting equal the coefficients of powers of β in the equation (11) we find the functions $u_0(\theta)$, $u_1(\theta)$, … . To obtain the values χ_0, χ_1, … we require the continuity of the functions $u_i(\theta)$ $(i = 0, 1, \dots)$ at the jump point. This requirement means that we glue the stable and the unstable integral manifolds at the jump point and, as a result, construct the canard passing through this point. As a result we have:

$$u_0(\theta) = \frac{(k_d e^{-\gamma\theta/2} + \chi_0 e^{\alpha_0\zeta E})\theta}{k_a e^{\gamma\theta/2}(1-\theta)}, \tag{12}$$

$$u_1(\theta) = \frac{-k_a u_0(\theta)(1-\theta)e^{\gamma\theta/2} + k_d e^{-\gamma\theta/2}\theta + 1 - u_0(\theta) + \chi_1 e^{\alpha_0\zeta E}\theta u_0'(\theta)}{k_a e^{\gamma\theta/2} u_0'(\theta)}, \tag{13}$$

$$\chi_0 = \frac{k_a(1-\bar\theta)e^{\gamma\bar\theta/2} - k_d e^{-\gamma\bar\theta/2}\bar\theta}{(k_a(1-\bar\theta)e^{\gamma\bar\theta/2} - 1)e^{\alpha_0\zeta E}\bar\theta}, \tag{14}$$

$$\chi_1 = -\frac{k_a u_1(\bar\theta)(1-\bar\theta)e^{\gamma\bar\theta/2} + u_1(\bar\theta) + k_a u_1(\bar\theta)u_1'(\bar\theta)(1-\bar\theta)e^{\gamma\bar\theta/2}}{e^{\alpha_0\zeta E}\bar\theta u_1'(\bar\theta)}, \tag{15}$$

where the value $\theta = \bar\theta$ corresponding to the jump point can be calculate from the system (5). The equations (12)–(15) define the first–order approximations for the canard passing through the jump point $(u(\bar\theta), \bar\theta)$ and the canard point of the system (1), (2). It should be noted that we can construct the canard either in the jump point A_1 (by gluing the stable slow integral manifold M_1 and the unstable one M_2, see Figures 9, 10) or in A_2 (by gluing the stable slow integral manifold M_3 and M_2, see Figures 7, 8). If it is necessary to glue stable and unstable slow invariant manifolds at the both jump points simultaneously, we should use two control parameters and as a result we obtain a canard cascade [12].

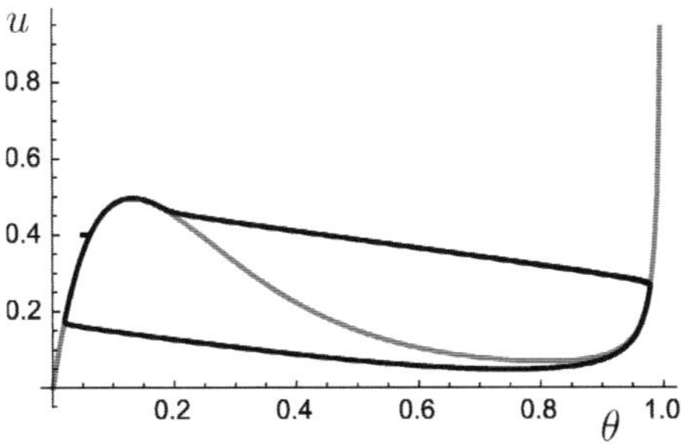

Figure 9. The slow curve (the red line) and the canard (the black line) of the system (1), (2); $k_e = 2.40855$.

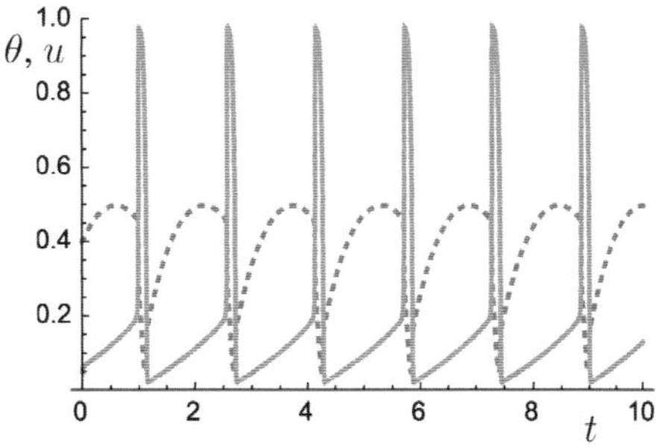

Figure 10. Plots of $\theta(t)$ (the solid orange line), $u(t)$ (the dashed blue line) for $k_e = 2.40855$.

4. Conclusion

We have explored the dynamic model of the electrocatalytic reaction under the influence of constantly present small perturbations. The critical regime separating the basic types of the regimes, slow and relaxation, was modelled with the help of the integral manifolds of variable stability. This approach was used in [12-21] for modelling of the critical phenomena in chemical systems.

We showed that in the differential system the critical regime corresponds to the phenomenon of the delayed loss of stability, when the trajectory, passing through a jump point, continues its motion for a while on the unstable slow manifold. The asymptotic formula for the critical value of the control parameter has been derived.

The obtained results is of utmost importance for several applications in chemical kinetics, as they can be used to determine the dynamics of the process in the chemical system for given initial conditions.

Acknowledgment
The contribution of N. Firstova was funded by RFBR and Samara Region according to the research project 16-41-630529 and the Ministry of Education and Science of the Russian Federation under the Competitiveness Enhancement Program of Samara University (2013-2020). E. Shchepakina was supported by the Ministry of Education and Science of the Russian Federation in the framework of the implementation of the Program "Research and development on priority directions of scientific-technological complex of Russia for 2014-2020" (Project RFMEFI58716X0033).

References

[1] Koper M T M and Sluyters J H 1994 Instabilities and oscillations in simple models of electrocatalytic surface reactions *Journal of Electroanalytical Chemistry* **371**(1) 149–159

[2] Berthier F, Diard J-P and Nugues S 1997 On the nature of the spontaneous oscillations observed for the Koper-Sluyters electrocatalitic reaction *Journal of Electroanalytical Chemistry* **436**(1) 35–42

[3] Mishchenko E F, Kolesov Yu S, Kolesov A Yu and Rozov N Kh 1995 *Asymptotic Methods in Singularly Perturbed Systems* (New York: Plenum Press)

[4] Shchepakina E and Sobolev V 2005 *Black Swans and Canards in Laser and Combustion Models* In: Singular Perturbation and Hysteresis (ed M P Mortell *et al*) (Philadelphia: SIAM) pp 207–255

[5] Shchepakina E, Sobolev V and Mortell M P 2014 *Singular Perturbations. Introduction to System Order Reduction Methods with Applications* In: Lecture Notes in Mathematics vol. 2114 (Cham–Heidelber–New York–Dordrecht–London: Springer)

[6] Firstova N and Shchepakina E 2016 Study of oscillatory processes in the one model of electrochemical reactor *CEUR Workshop Proceedings* **1638** 731–741

[7] Benoit E, Callot J L, Diener F and Diener M 1981–1982 Chasse au canard *Collectanea Mathematica* **31–32** 37–119 [in French]

[8] Shchepakina E and Korotkova O 2011 Condition for canard explosion in a semiconductor optical amplifier *Journal of the Optical Society of America B: Optical Physics* **28**(8) 1988–93

[9] Shchepakina E and Korotkova O 2013 Canard explosion in chemical and optical systems *Discrete and Continuous Dynamical Systems - Series B* **18**(2) 495–512

[10] Pokrovskii A, Rachinskii D, Sobolev V and Zhezherun A 2011 Topological degree in analysis of canard-type trajectories in 3-D systems *Applicable Analysis* **90**(7) 1123–39

[11] Shchepakina E and Sobolev V 2016 Invariant surfaces of variable stability *Journal of Physics: Conference Series* **727** 012016

[12] Sobolev V 2013 Canard Cascades *Discrete and Continuous Dynamical Systems B* **18**(3) 513–521

[13] Gorelov G N and Sobolev V A 1991 Mathematical modelling of critical phenomena in thermal explosion theory *Combustion and Flame* **87** 203–210

[14] Gorelov G N and Sobolev V A 1992 Duck–trajectories in a thermal explosion problem *Applied Mathematics Letters* **5** 3–6

[15] Shchepakina E 2001 Critical conditions of self-ignition in dusty media *Journal of Advances in Chemical Physics* **20**(7) 3–9

[16] Shchepakina E A 2002 Attracting/repelling integral surfaces in combustion problems *Mathematical Models and Computer Simulations* **14**(3) 30–42 [in Russian]

[17] Shchepakina E A 2003 Black swans and canards in self-ignition problem *Nonlinear Analysis: Real World Appllications* **4**(1) 45–50

[18] Shchepakina E 2003 Singular perturbations in the problem of modeling of safe combustion regimes *Mathematical Models and Computer Simulations* **15**(8) 113–117 [in Russian]

[19] Shchepakina E 2005 Canards and black swans in model of a 3-D autocatalator *Journal of Physics: Conference Series* **22** 194–207

[20] Golodova E S and Shchepakina E A 2009 Modeling of safe combustion at the maximum temperature *Mathematical Models and Computer Simulations* **1**(2) 322–334

[21] Shchepakina E A 2015 Critical phenomena in a model of fuel's heating in a porous medium *CEUR Workshop Proceedings* **1490** 179–189

IOP Conf. Series: Journal of Physics: Conf. Series **811** (2017) 012003 doi:10.1088/1742-6596/811/1/012003

Weak periodic solutions of $x\ddot{x}+1=0$ and the Harmonic Balance Method

J D García-Saldaña[1,3] **and A Gasull** [2]

[1,2] Departament de Matemàtiques, Universitat Autònoma de Barcelona
Edifici C. 08193 Bellaterra, Barcelona. Spain.

E-mail: `jgarcias@ucsc.cl, gasull@mat.uab.cat`

Abstract. We prove that the differential equation $x\ddot{x}+1=0$ has continuous weak periodic solutions and compute their periods. Then, we use the Harmonic Balance Method until order six to approximate these periods and to illustrate how the accuracy of the method increases with the order. Our computations rely on the Gröbner basis approach.

1. Introduction and main results

The nonlinear differential equation

$$x\ddot{x}+1=0 \tag{1}$$

appears in the modeling of certain phenomena in plasma physics. More concretely, studying a certain electron beam injected into a plasma tube, where the magnetic field is cylindrical and increases towards the axis in inverse proportion to the radius, see [1, 8]. In [11] and [13, Sec. 3.2.2], the author calculates the period of its periodic orbits and use the N-th order Harmonic Balance Method (HBM), for $N=1,2$, to obtain approximations of these periodic solutions and of their corresponding periods. Similarly, the same functions are approximated in [8]. However, strictly speaking, it can be seen that neither equation (1), nor its associated planar system

$$\begin{cases} \dot{x}=y, \\ \dot{y}=-\frac{1}{x}, \end{cases} \tag{2}$$

which is singular at $x=0$, have smooth periodic solutions. For system (2) this is so, because it has no critical points and it is well-known ([9]) that periodic orbits of planar autonomous systems must surround a critical point. For equation (1) it is not difficult to see that any periodic solution $x(t)$ must vanish for some $t^*\in\mathbb{R}$, that is $x(t^*)=0$. Then $\lim_{t\to t^*}\ddot{x}(t)=\infty$ and as a consequence this equation cannot have $\mathcal{C}^2$-periodic solutions.

We start giving two different interpretations of the computations of [8, 11] of the period function. The first one states their results in terms of weak (or generalized) solutions, where in this work a weak solution is a function satisfying the differential equation (1) on an open and dense set, but which is only continuous ($\mathcal{C}^0$) at some isolated points.

[3] Current address: Departamento de Matemática y Física Aplicadas, Facultad de Ingeniería, Universidad Católica de la Santísima Concepción. Casilla 297, Concepción, Chile.

The second interpretation shows that, for a given initial condition, this period is the limit when k tends to zero, of the period of an actual periodic solution with the same initial condition, of the smooth planar differential system

$$\begin{cases} \dot{x} = y, \\ \dot{y} = -\frac{x}{x^2+k^2}, \quad k \neq 0, \end{cases} \tag{3}$$

which has a global center at the origin.

Next theorem collects our first results. The method used to obtain $T(A)$ is based on standard procedures applied to 1-dimensional oscillatory physical systems, see [2, 12, 13, 15, 16, 19], adapted to this singular setting. Item (ii) provides a new result. It is a consequence of applying the Lebesgue's dominated convergence theorem to the integral expression that gives the periods of the periodic orbits of system (3).

Theorem 1.1. *(i) For the initial conditions $x(0) = A$, $\dot{x}(0) = 0$, with $A \neq 0$, the differential equation (1) has a weak $\mathcal{C}^0$-periodic solution with period $T(A) = 2\sqrt{2\pi}A$.*

(ii) Let $T(A;k)$ be the period of the periodic orbit of system (3) with initial conditions $x(0) = A$, $y(0) = 0$. Then

$$T(A;k) = 4\,A \int_0^1 \frac{ds}{\sqrt{\ln\left(\frac{A^2+k^2}{A^2s^2+k^2}\right)}}$$

and

$$\lim_{k\to 0} T(A;k) = 4A \int_0^1 \frac{1}{\sqrt{-2\ln s}} ds = 2\sqrt{2\pi}A = T(A).$$

Recall that the N-th order HBM consists in approximating the periodic solutions of differential equations by truncated Fourier series with N harmonics and an unknown frequency; see for instance [6, 7, 12, 13] or Section 3 for a short overview of the method. In [13] the author asks for techniques to deal analytically with the N-th order HBM, for $N \geq 3$. In [5] it is shown how resultants can be used when $N = 3$.

Next, in this work we will show how to utilize a more powerful tool, the computation of Gröbner basis ([3, Ch. 5]), to obtain higher accuracy in the approximations obtained by using the N-th order HBM, for $N \geq 3$. As an example, we will apply this method to approach the function $T(A)$ introduced in Theorem 1.1. Clearly, it can also be applied to other similar differential equations to approach their periodic orbits and their corresponding periods, like for instance, the ones considered in [2, 14, 15, 16]

Notice that all solutions of equation (1) are also solutions of the family of differential equations

$$x^{m+1}\ddot{x} + x^m = 0, \quad m \in \mathbb{N} \cup \{0\}. \tag{4}$$

In fact, in this new family only the new solution $x = 0$ is added to the solutions of (1).

In the second part of this paper we address the question of the effect of this parameter m and of the integer N, that measures the order of the HBM, on the accuracy of the approaches to the period function,

$$T(A) = 2\sqrt{2\pi}A \approx 5.0132A,$$

by the periods of the trigonometric polynomials obtained from applying the N-th order HBM to (4). Our results are given in next theorem, where $[a]$ denotes the integer part of a. In particular, it can be seen that at first order, the best approximations of $T(A)$ are obtained when $m \in \{1, 2\}$ and not when $m = 0$.

IOP Conf. Series: Journal of Physics: Conf. Series **811** (2017) 012003 doi:10.1088/1742-6596/811/1/012003

Theorem 1.2. *Let $\mathcal{T}_N(A;m)$ be the period of the truncated Fourier series obtained from applying the N-th order HBM to equation (4). Then there exist constants $C_N(m)$ such that $\mathcal{T}_N(A;m) = C_N(m)A$. Moreover*

(i) For all $m \in \mathbb{N} \cup \{0\}$,

$$C_1(m) = 2\pi\sqrt{\frac{2[\frac{m+1}{2}]+1}{2[\frac{m+1}{2}]+2}}. \tag{5}$$

(ii) For $m = 0$,

$$C_1(0)=\sqrt{2}\,\pi \approx 4.4428, \qquad C_4(0) \approx 5.0455,$$

$$C_2(0)=(\sqrt{218}/9)\,\pi \approx 5.1539, \qquad C_5(0) \approx 4.9841,$$

$$C_3(0)=\frac{13810534\,\pi}{3\sqrt{5494790257313+115642506449\sqrt{715}}} \approx 4.9353, \qquad C_6(0) \approx 5.0260.$$

(iii) For $m = 1$,

$$C_1(1) = \sqrt{3}\,\pi \approx 5.4414, \qquad C_3(1) \approx 5.1476,$$

$$C_2(1) \approx 5.2733, \qquad C_4(1) \approx 5.1186.$$

(iv) For $m = 2$,

$$C_1(2) = \sqrt{3}\,\pi \approx 5.4414, \qquad C_2(2) \approx 5.2724, \qquad C_3(2) \approx 5.1417.$$

Moreover, the approximate values stated above are the roots of certain given polynomials with integer coefficients, and they can be obtained with any desired accuracy using Sturm sequences.

Notice that the values $C_1(m)$, for $m \in \{0, 1, 2\}$ given in items (ii), (iii) and (iv), respectively, are already computed in item (i). We only make them explicit to clarify the reading.

Table 1. Percentage of relative errors $e_N(m)$.

$e_N(m)$	$m = 0$	$m = 1$	$m = 2$
$N = 1$	11.38%	8.54%	8.54%
$N = 2$	2.80%	5.19%	5.17%
$N = 3$	1.55%	2.68%	2.56%
$N = 4$	0.64%	2.10%	–
$N = 5$	0.58%	–	–
$N = 6$	0.25%	–	–

To elucidate which of the approaches given in Theorem 1.2 is better, in Table 1 we give the percentage of the relative errors

$$e_N(m) = 100\left|\frac{\mathcal{T}_N(A;m) - T(A)}{T(A)}\right| = 100\left|\frac{C_N(m) - 2\sqrt{2\pi}}{2\sqrt{2\pi}}\right|.$$

IOP Conf. Series: Journal of Physics: Conf. Series **811** (2017) 012003 doi:10.1088/1742-6596/811/1/012003

The best approximation that we have found corresponds to $\mathcal{T}_6(A;0)$. Our computers have not been able to get the Gröbner basis needed to fill the gaps of the table.

The paper is organized as follows. Theorem 1.1 is proved in Section 2. In Section 3 we describe the N-th order HBM adapted to our purpose. Finally, in Section 4, we use this method to prove Theorem 1.2.

2. Proof of Theorem 1.1.

(i) We start by proving that the solution of (1) with initial conditions $x(0) = A$, $\dot{x}(0) = 0$ and for $t \in \left(-\frac{\sqrt{2\pi}}{2}A, \frac{\sqrt{2\pi}}{2}A\right)$ is

$$x(t) = \phi_0(t) := Ae^{-\left(\operatorname{erf}^{-1}\left(\frac{2t}{\sqrt{2\pi}A}\right)\right)^2}, \tag{6}$$

where erf^{-1} is the inverse of the error function

$$\operatorname{erf}(z) = \frac{2}{\sqrt{\pi}} \int_0^z e^{-s^2}\, ds.$$

Notice that $\lim_{t\to\pm\frac{\sqrt{2\pi}}{2}} \phi_0(t) = 0$ and $\lim_{t\to\pm\frac{\sqrt{2\pi}}{2}} \phi_0'(t) = \mp\infty$. To obtain (6), observe that from system (2) we arrive at the simple differential equation $\frac{dx}{dy} = -xy$, which has separable variables and can be solved by integration. The particular solution that passes through the point $(x, y) = (A, 0)$ is $x = Ae^{-y^2/2}$. Plugging this equation in (2) we get $\frac{dy}{dt} = -\frac{e^{y^2/2}}{A}$, which is again a separable equation. It has the solution

$$y(t) = -\sqrt{2}\,\operatorname{erf}^{-1}\left(\frac{2t}{\sqrt{2\pi}A}\right), \tag{7}$$

which is well defined for $t \in \left(-\frac{\sqrt{2\pi}}{2}A, \frac{\sqrt{2\pi}}{2}A\right)$ since $\operatorname{erf}^{-1}(\cdot)$ is defined in $(-1, 1)$. Finally, since $x(t) = Ae^{-y^2(t)/2}$, we get (6).

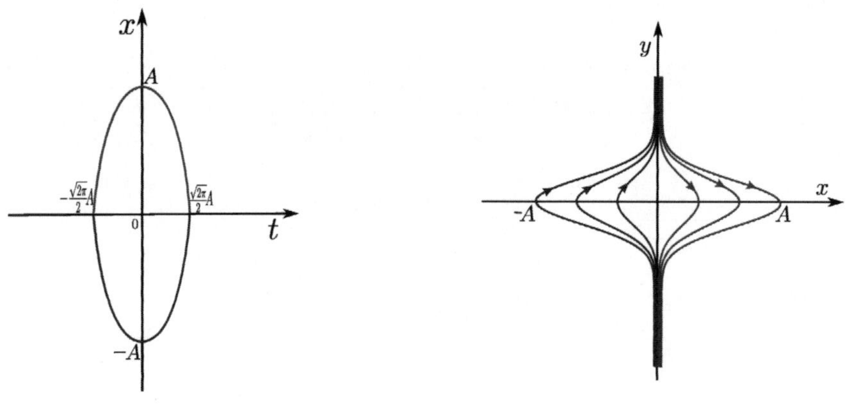

Figure 1. (a) Two solutions of equation (1) (b) Phase-portrait of system (2)

By using $x(t)$ and $y(t)$ given by (6) and (7), respectively, we can draw the phase portrait of (2) which, as we can see in Figure 1.(b), is symmetric with respect to both axes. Notice that its orbits do not cross the y-axis, which is a singular locus for the associated vector field. Moreover,

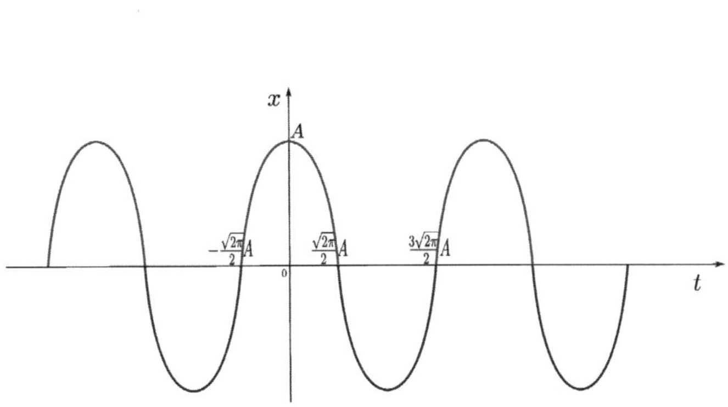

Figure 2. A weak $\mathcal{C}^0$-periodic solution of (1).

the solutions of (1) are not periodic (see Figure 1.(a)), and the transit time of $x(t)$ from $x = A$ to $x = 0$ is $\sqrt{2\pi}\,A/2$.

From (6) we introduce the $\mathcal{C}^0$-function, defined on $\mathbb{R}$, as

$$\phi(t) = \begin{cases} (-1)^n \phi_0(t - n\sqrt{2\pi}), & \text{for } \; t \in \left(\frac{2n-1}{2}\sqrt{2\pi}, \frac{2n+1}{2}\sqrt{2\pi}\right), \quad n \in \mathbb{Z}, \\ 0, & \text{for } \; t = \frac{2n+1}{2}\sqrt{2\pi}, \qquad n \in \mathbb{Z}, \end{cases}$$

see Figure 2. It is a $\mathcal{C}^0$-periodic function of period $T(A) = 2\sqrt{2\pi}A$ and $x = \phi(t)$ satisfies (1) for all $t \in \mathbb{R} \setminus \cup_{n\in\mathbb{Z}}\{\frac{2n+1}{2}\sqrt{2\pi}\}$. Hence (i) of the theorem follows.

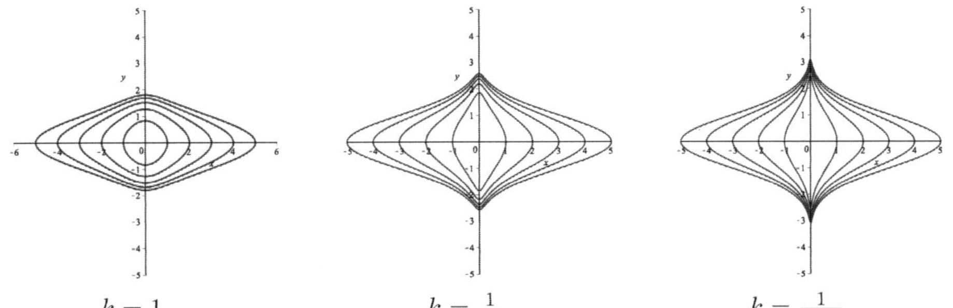

$k = 1$ $k = \frac{1}{50}$ $k = \frac{1}{1000}$

Figure 3. Phase portraits of (3) for different values of k.

(ii) System (3) is Hamiltonian with Hamiltonian function

$$H(x, y) = \frac{y^2}{2} + \frac{\ln(x^2 + k^2)}{2}.$$

Since $\ln(x^2 + k^2)$ has a global minimum at 0 and $\ln(x^2 + k^2)$ tends to infinity when $|x|$ does, system (3) has a global center at the origin. In Figure 3 we can see its phase portrait for some values of k. This figure also illustrates how the periodic orbits of (3) approach to the solutions of system (2).

Its period function is

$$T(A; k) = 2\int_{-A}^{A} \frac{dx}{y(x)} = 2\int_{-A}^{A} \frac{dx}{\sqrt{2h - \ln(x^2 + k^2)}},$$

where $h = \ln(A^2 + k^2)/2$ is the energy level of the orbit passing through the point $(A, 0)$. Therefore,

$$T(A;k) = 2\int_{-A}^{A} \frac{dx}{\sqrt{\ln\left(\frac{A^2+k^2}{x^2+k^2}\right)}} = 4\,A\int_0^1 \frac{ds}{\sqrt{\ln\left(\frac{A^2+k^2}{A^2s^2+k^2}\right)}},$$

where we have used the change of variable $s = x/A$ and the symmetry with respect to x. Then,

$$\lim_{k\to 0} T(A;k) = \lim_{k\to 0}\int_0^1 \frac{4\,A\,ds}{\sqrt{\ln\left(\frac{A^2+k^2}{A^2s^2+k^2}\right)}}.$$

If we prove that

$$\lim_{k\to 0}\int_0^1 \frac{4\,A\,ds}{\sqrt{\ln\left(\frac{A^2+k^2}{A^2s^2+k^2}\right)}} = \int_0^1 \lim_{k\to 0}\frac{4\,A\,ds}{\sqrt{\ln\left(\frac{A^2+k^2}{A^2s^2+k^2}\right)}}, \tag{8}$$

then

$$\lim_{k\to 0} T(A;k) = 4\,A\int_0^1 \frac{ds}{\sqrt{-2\ln(s)}} = 2\sqrt{2\pi}A = T(A)$$

and the theorem will follow. To show that (8) holds, take any sequence $1/z_n$, with z_n tending monotonically to infinity, and consider the functions $f_n(s) = \left(\ln\left(\frac{(Az_n)^2+1}{(Az_n)^2s^2+1}\right)\right)^{-1/2}$. We have that the sequence $\{f_n(s)\}_{n\in\mathbb{N}}$ is formed by measurable and positive functions defined on the interval $(0, 1)$. It is not difficult to prove that it is a decreasing sequence. In particular, $f_n(s) < f_1(s)$ for all $n > 1$. Therefore, if we show that $f_1(s)$ is integrable, then we can apply the Lebesgue's dominated convergence theorem ([17]) and (8) will follow. The function $f_1(s)$ tends to infinity when $s \to 1^-$. Hence, to prove that $\int_0^1 f_1(s)\,ds < \infty$ we need to study $f_1(s)$ near $s = 1$. Set $B = (Az_1)^{-2}$ and $s = 1 - w$, with $w > 0$ and $w \sim 0$. Then,

$$\frac{(Az_1)^2+1}{(Az_1)^2s^2+1} = \frac{B^{-1}+1}{B^{-1}s^2+1} = \frac{1+B}{(1-w)^2+B} = \frac{1+B}{1+B-2w+w^2} = \frac{1}{1-\frac{2}{1+B}w+O(w^2)} =$$
$$= 1 + \frac{2}{1+B}w + O(w^2) = 1 + \frac{2(Az_1)^2}{(Az_1)^2+1}(1-s) + O((1-s)^2).$$

By using that when $x \sim 0$, $\ln(1 + x) = x + O(x^2)$, it holds that when $s < 1$, $s \sim 1$,

$$f_1(s) = \left(\ln\left(\frac{A^2z_1^2+1}{A^2z_1^2s^2+1}\right)\right)^{-1/2} = \left(\ln\left(1 + \frac{2(Az_1)^2}{(Az_1)^2+1}(1-s) + O((1-s)^2)\right)\right)^{-1/2}$$
$$\sim \left(\frac{2(Az_1)^2(1-s)}{(Az_1)^2+1}\right)^{-1/2}.$$

Since this last expression is integrable, the convergence of the integral is consequence of the comparison test for improper integrals.

3. The Harmonic Balance Method

This section gives a brief description of the HBM applied to second order differential equations

$$\mathcal{F} := \mathcal{F}(x(t), \ddot{x}(t)) = 0, \tag{9}$$

with $\mathcal{F}:\mathbb{R}^2\to\mathbb{R}$ a smooth function satisfying $\mathcal{F}(-u,-v)=\mathcal{F}(u,v)$, because of our interests. Notice that if $x(t)$ is a solution of (9) then $x(-t)$ is also a solution.

Suppose that equation (9) has a T-periodic solution $x(t)$ with initial conditions $x(0)=A$, $\dot{x}(0)=0$ and period $T=T(A)$. If $x(t)$ satisfies $x(t)=x(-t)$ it is clear that its Fourier series writes as

$$\sum_{k=1}^{\infty} a_k\cos(k\omega t),\quad \text{with}\quad \sum_{k=1}^{\infty} a_k = A \quad\text{and}\quad \omega=\frac{2\pi}{T}.$$

As we have seen in the previous section, the weak periodic solutions of equation (1) which we want to approximate satisfy the above property. Moreover, $x(T/4)=0$ and $\dot{x}(T/4)$ does not exist. In any case, smooth approximations to $x(t)$, should also satisfy $\dot{x}(t)\ddot{x}(t)+x(t)\dddot{x}(t)=0$, and hence $\dot{x}(T/4)=0$. For this reason, in this work we will consider Fourier series in cosine, not including the even terms $\cos(2j\omega t)$, $j\in\mathbb{N}\cup\{0\}$, which do not satisfy this property. This type of a priori simplifications are similar to the ones introduced in [10] for other problems.

Hence, in our setting, the HBM of order N follows the next five steps:

1. Consider a trigonometric polynomial

$$x_N(t)=\sum_{j=1}^{N} a_{2j-1}\cos((2j-1)\omega_N t)\quad\text{with}\quad \sum_{j=1}^{N} a_{2j-1}=A. \tag{10}$$

2. Compute the $2\pi/\omega_N$-periodic function $\mathcal{F}_N := \mathcal{F}(x_N(t),\ddot{x}_N(t))$, which also has an associated Fourier series,

$$\mathcal{F}_N(t)=\sum_{j\ge 0}\mathcal{A}_j\cos(j\,\omega_N t),$$

where $\mathcal{A}_j=\mathcal{A}_j(\mathbf{a},\omega_N,A)$ $j\ge 0$, with $\mathbf{a}=(a_1,a_3,\ldots,a_{2N-1})$.

3. Find all values $\mathbf{a}$ and ω_N such that

$$\mathcal{A}_j(\mathbf{a},\omega_N,A)=0 \qquad \text{for}\quad 1\le j\le j_N, \tag{11}$$

where j_N is the value such that (11) consists exactly of N non trivial equations. Notice also that each equation $\mathcal{A}_j(\mathbf{a},\omega_N,A)=0$ is equivalent to

$$\int_0^{2\pi/\omega_N}\cos(j\omega_N t)\mathcal{F}_N(t)\,dt=0. \tag{12}$$

4. Then the expression (10), with the values of $\mathbf{a}=\mathbf{a}(A)$ and $\omega_N=\omega_N(A)$ obtained in point *3*, provides candidates for the approximations of the actual periodic solutions of the initial differential equation. In particular, the functions $\mathcal{T}_N=\mathcal{T}_N(A)=2\pi/\omega_N$ give approximations of the periods of the corresponding periodic orbits.

5. The final approximation is chosen to be the one associated to the solution that minimizes the norm

$$||\mathcal{F}_N(t)||=\int_0^{\mathcal{T}_N}\mathcal{F}_N^2(t)\,dt,$$

because an actual solution of our differential equation would satisfy $||\mathcal{F}(x(t),\ddot{x}(t))||=0$ and $\mathcal{F}_N(t)=\mathcal{F}(x_N(t),\ddot{x}_N(t))$ must approach $\mathcal{F}(x(t),\ddot{x}(t))$.

Remark 3.1. *(i) Notice that going from order N to order $N+1$ in the method implies to compute again all the coefficients of the Fourier polynomial, because in general the coefficients of the Fourier series of $x_N(t)$ and $x_{N+1}(t)$ do not coincide.*

(ii) The above set of equations (11) is a system of polynomial equations which usually is not easy to solve. For this reason only the values of $N=1,2$ are considered in many works, see

for instance [11, 13] *and the references therein. To solve system* (11) *for* $N \geq 3$ *we use the Gröbner basis approach* ([3]). *In general this method is faster that using successive resultants and moreover it does not give spurious solutions.*

(iii) As far as we know, the test proposed in point 5 to select the best approach is not commonly used. We propose it following the definition of accuracy of an approximated solution used in [4] *and inspired in the classical works* [18, 20].

4. Application of the HBM

We start proving a lemma that allows to reduce our computations to the case $A = 1$.

Lemma 4.1. *Let* $\mathcal{T}_N(A; m)$ *be the period of the truncated Fourier series obtained from applying the* N*-th order HBM to equation* (4). *Then there exist constants* $C_N(m)$ *such that* $\mathcal{T}_N(A; m) = C_N(m)A$.

Proof. Consider $\mathcal{F}_N = x_N^{m+1}\ddot{x}_N + x_N^m = 0$, with x_N given in (10). We have to solve the set of $N+1$ non-trivial equations

$$\int_0^{2\pi/\omega_N} \cos(j\omega_N t)\mathcal{F}_N(t)\,dt = 0 \qquad 1 \leq j \leq j_N, \qquad \sum_{j=1}^{N} a_{2j-1} = A, \tag{13}$$

with $N+1$ unknowns $a_1, a_3, \ldots, a_{2N-1}$ and ω_N and $A \neq 0$. The lemma clearly follows if we prove the following assertion: $\tilde{a}_1, \tilde{a}_3, \ldots, \tilde{a}_{2N-1}$ and $\tilde{\omega}_N$ is a solution of (13) with $A = 1$ if and only if $A\tilde{a}_1, A\tilde{a}_3, \ldots, A\tilde{a}_{2N-1}$ and $\tilde{\omega}_N/A$ is a solution of (13). This equivalence is a consequence of the fact that the change of variables $s = At$ transforms the integral equation (13) into

$$\frac{1}{A}\int_0^{2\pi A/\omega_N} \cos\Big(j\frac{\omega_N}{A}s\Big)\mathcal{F}_N\Big(\frac{s}{A}\Big)\,ds = 0$$

and of the structure of the right-hand side equation of (13).

Hence, $\mathcal{T}_N(A; m) = \mathcal{T}_N(1; m)A =: C_N(m)A$, as we wanted to prove. □

Proof of Theorem 1.2. By Lemma 4.1 we know that $\mathcal{T}_N(A; m) = C_N(m)A$ for some unknown constants $C_N(m)$. Therefore we can restrict our attention to the case $A = 1$.

(i) Following section 3, we consider $x_1(t) = \cos(\omega_1 t)$ as the first approximation to the actual solution of the functional equation $\mathcal{F}(x(t), \ddot{x}(t)) = x^{m+1}\ddot{x} + x^m = 0$. Then

$$\mathcal{F}_1(t) = -\omega_1^2 \cos^{m+2}(\omega_1 t) + \cos^m(\omega_1 t).$$

When $m = 2k$ the above expression writes as

$$\mathcal{F}_1(t) = -\omega_1^2 \cos^{2k+2}(\omega_1 t) + \cos^{2k}(\omega_1 t) = 0.$$

Using (12) for $j = 0$ we get

$$\int_0^{2\pi/\omega_1} \mathcal{F}_1(t)\,dt = -\omega_1^2 I_{2k+2} + I_{2k} = 0, \tag{14}$$

where $I_{2\ell} = \int_0^{2\pi/\omega_1} \cos^{2\ell}(\omega_1 t)dt$. Using integration by parts we prove that $(2k+2)I_{2k+2} = (2k+1)I_{2k}$. Combining this equality and (14) we obtain

$$\omega_1 = \sqrt{\frac{2k+2}{2k+1}},$$

or equivalently,

$$C_1(m) = 2\pi\sqrt{\frac{2k+1}{2k+2}},$$

which in terms of m coincides with (5). The case m odd follows similarly. The only difference is that to find $\mathcal{T}_1(A;m)$, instead of condition (14) we have to impose that

$$\int_0^{2\pi/\omega_1} \cos(\omega_1 t)\mathcal{F}_1(t)\,dt = 0,$$

because $\int_0^{2\pi/\omega_1} \mathcal{F}_1(t)\,dt \equiv 0$.

(*ii*) Case $m = 0$. Consider the functional equation $\mathcal{F}(x(t), \ddot{x}(t)) = x(t)\ddot{x}(t) + 1 = 0$.

When $N = 2$, we take as approximation $x_2(t) = a_1\cos(\omega_2 t) + a_3\cos(3\omega_2 t)$. The vanishing of the coefficients of 1 and $\cos(2\omega_2 t)$ in the Fourier series of $\mathcal{F}_2$ provides the nonlinear system

$$\begin{aligned} 1 - \frac{1}{2}(a_1^2 + 9a_3^2)\omega_2^2 &= 0, \\ a_1 + 10a_3 &= 0, \\ a_1 + a_3 - 1 &= 0. \end{aligned}$$

By solving it and applying point *5* of the HBM we get that $\omega_2 = 18/\sqrt{218}$. Therefore,

$$C_2(0) = \frac{\sqrt{218}}{9}\pi \approx 5.1539,$$

as we wanted to prove.

For the third-order HBM we use $x_3(t) = a_1\cos(\omega_3 t) + a_3\cos(3\omega_3 t) + a_5\cos(5\omega_3 t)$ as the approximate solution. Imposing that the coefficients of 1, $\cos(2\omega_3 t)$, and $\cos(4\omega_3 t)$ in $\mathcal{F}_3$ vanish we arrive at the system

$$\begin{aligned} P &= 2 - \left(a_1^2 + 9a_3^2 + 25a_5^2\right)\omega_3^2 = 0, \\ Q &= a_1^2 + 10a_1a_3 + 34a_3a_5 = 0, \\ R &= 5a_3 + 13a_5 = 0, \\ S &= a_1 + a_3 + a_5 - 1 = 0. \end{aligned}$$

Since all the equations are polynomial, the solutions can be found by using the Gröbner basis approach, see [3]. Recall that the idea of this approach consists in finding a new system of generators, say $G_1, G_2, \ldots, G_\ell$, of the ideal of $\mathbb{R}[a_1, a_3, a_5, \omega_3]$ generated by P, Q, R and S. Hence, solving $P = Q = R = S = 0$ is equivalent to solving $G_i = 0$, $i = 1, \ldots, \ell$. In general, choosing the lexicographic order in the Gröbner basis approach, we get that the polynomials of the equivalent system have triangular structure with respect to the variables and it can be easily solved.

Now, by computing the Gröbner basis of $\{P, Q, R, S\}$ with respect to the lexicographic order $[a_1, a_3, a_5, \omega_3]$ we obtain a new basis with four polynomials ($\ell = 4$), one of them being,

$$G_1(\omega_3) = 1553685075\omega_3^8 - 3692301106\omega_3^6 + 2143547654\omega_3^4 - 402413472\omega_3^2 + 20301192.$$

Solving $G_1(\omega_3) = 0$ and using again point *5* of our approach to HBM we get that the solution which yields the best approximation gives

$$\omega_3 = \frac{3\sqrt{5494790257313 + 115642506449\sqrt{715}}}{6905267}.$$

Hence, the expression $C_3(0) = 2\pi/\omega_3$ of the statement follows.

When $N = 4$ we consider $x_4(t) = a_1 \cos(\omega_4 t) + a_3 \cos(3\omega_4 t) + a_5 \cos(5\omega_4 t) + a_7 \cos(7\omega_4 t)$, and we arrive at the system

$$\begin{aligned} P &= 2 - \left(a_1^2 + 9a_3^2 + 25a_5^2 + 49a_7^2\right)\omega_4^2 = 0, \\ Q &= a_1^2 + 10a_1a_3 + 34a_3a_5 + 74a_5a_7 = 0, \\ R &= 5a_1a_3 + 13a_1a_5 + 29a_3a_7 = 0, \\ S &= 9a_3^2 + 50a_1a_7 + 26a_1a_5 = 0, \\ U &= a_1 + a_3 + a_5 + a_7 - 1 = 0. \end{aligned}$$

The Gröbner basis of $\{P, Q, R, S, U\}$ with respect to the order $[a_1, a_3, a_5, a_7, \omega_4]$ is a new basis with five polynomials, one of them being an even polynomial in ω_4 of degree 16 with integers coefficients. Solving it we obtain that the best approximation is $\omega_4 \approx 1.2453$, which gives $C_4(0) \approx 5.0455$.

For $N = 5$ and $N = 6$ we have done similar computations. In the case $N = 5$ one of the generators of the Gröbner basis is an even polynomial in ω_5 with integers coefficients and degree 32. When $N = 6$ the same happens but with a polynomial of degree 64 in ω_6. Solving the corresponding polynomials we get that $\omega_5 \approx 1.2606$ and $\omega_6 \approx 1.2501$, and consequently, $C_5(0) \approx 4.9843$, and $C_6(0) \approx 5.0260$.

(iii) Case $m = 1$. We apply the HBM to $\mathcal{F}(x(t), \ddot{x}(t)) = x^2(t)\ddot{x}(t) + x(t) = 0$.

When $N = 2$, doing similar computations as in item *(ii)*, we arrive at

$$\begin{aligned} P &= 4 - \left(3a_1^2 + 11a_1a_3 + 38a_3^2\right)\omega_2^2 = 0, \\ Q &= 4a_3 - \left(a_1^3 + 22a_1^2a_3 + 27a_3^3\right)\omega_2^2 = 0, \\ R &= a_1 + a_3 - 1 = 0. \end{aligned}$$

Again, by computing the Gröbner basis of $\{P, Q, R\}$ with respect to the lexicographic order $[a_1, a_3, \omega_2]$ we obtain a new basis with three polynomials, one of them being

$$G_1(\omega_2) = 7635411\omega_2^8 - 14625556\omega_2^6 + 5833600\omega_2^4 - 661376\omega_2^2 + 13824.$$

Notice that the equation $G_1(\omega_2) = 0$ can be algebraically solved. Nevertheless, for the sake of shortness, we do not give the exact roots. Following again step *5* of our approach we get that the best solution is $\omega_2 \approx 1.1915$, or equivalently that $C_2(1) \approx 5.2733$.

The HBM when $N = 3$ produces the system

$$\begin{aligned} P &= 4a_1 - \left(3a_1^3 + 11a_1^2a_3 + 38a_1a_3^2 + 70a_1a_3a_5 + 102a_1a_5^2 + 43a_3^2a_5\right)\omega_3^2 = 0, \\ Q &= 4a_3 - \left(a_1^3 + 22a_1^2a_3 + 27a_1^2a_5 + 70a_1a_3a_5 + 27a_3^3\right)\omega_3^2 = 0, \\ R &= 4a_5 - \left(11a_1^2a_3 + 54a_1^2a_5 + 19a_1a_3^2 + 86a_3^2a_5 + 75a_5^3\right)\omega_3^2 = 0, \\ S &= a_1 + a_3 + a_5 - 1 = 0. \end{aligned}$$

Computing the Gröbner basis of $\{P, Q, R, S\}$ with respect to the lexicographic order $[a_1, a_3, a_5, \omega_3]$ we get that one of the polynomials of the new basis is an even polynomial in ω_3 of degree 26 with integer coefficients. By solving it we obtain that the best approximation is $\omega_3 \approx 1.2206$, which produces the value $C_3(1)$ of the statement.

When $N = 4$ we arrive at five polynomial equations, which we omit. Once more, using the Gröbner basis approach we obtain a polynomial condition in ω_4 of degree 80. Finally, $\omega_4 \approx 1.2275$ and $C_4(1) \approx 5.1186$.

(*iv*) When $m = 2$ we have to approximate the solutions of $\mathcal{F}(x(t), \ddot{x}(t)) = x^3(t)\ddot{x}(t) + x^2(t) = 0$. We do not give the details of the proof because the results are obtained by using exactly the same type of computations as above.

□

Remark 4.2. *For each N and m our computations also provide a trigonometric polynomial that approximates the continuous weak periodic solution $\phi(t)$ given in the proof of Theorem 1.1.*

Acknowledgments

The first author is supported by FONDECYT Postdoctoral fellowship 3150131/2015 and was supported by the grant FPU AP2009-1189. The second author is supported by a MINECO/FEDER grant number MTM2016-77278-P and by a CIRIT grant number 2014SGR568.

References

[1] Acton J R and Squire P T 1985 *Solving Equations with Physical Understanding* ed Adam Hilger (Bristol and Boston) 219 pp
[2] Beléndez A, Gimeno E, Álvarez M L, Gallego S, Ortuño M and Méndez D I 2009 A novel rational harmonic balance approach for periodic solutions of conservative nonlinear oscillators *International Journal of Nonlinear Sciences & Numerical Simulation* **10** 13–26.
[3] Cox D, Little J and O'Shea D 2007 *Ideals, Varieties, and Algorithms: An Introduction to Computational Algebraic Geometry and Commutative Algebra* Third edition Undergraduate Texts in Mathematics (New York: Springer) 551 pp
[4] García-Saldaña J D and Gasull A 2013 A theoretical basis for the harmonic balance method *J. Differ Equations* **254** 67–80
[5] García-Saldaña J D and Gasull A 2015 The period function and the harmonic balance method *Bull. Sci. Math.* **139** 33–60
[6] Giné J 2013 On the determination of the limit cycles using the harmonic balance method *J. Math. Phys.* **54** 103510
[7] Guo Z and Leung A Y T 2010 The iterative homotopy harmonic balance method for conservative Helmholtz-Duffing oscillators *Appl. Math. Comput.* **215** 3163–69
[8] He J-H 2001 An analytic solution to certain nonlinear equations in plasma physics *Ukr. J. Phys.* **46** 567–568
[9] Lefschetz S 1977 *Differential Equations: Geometric Theory* reprinting of the second edition (New York: Dover Publications Inc.) 390 pp
[10] Mickens R E 2002 Fourier representations for periodic solutions of odd-parity systems *J. Sound Vibration* **258** 398–401
[11] Mickens R E 2007 Harmonic balance and iteration calculations of periodic solutions to $\ddot{y} + y^{-1} = 0$ *J. Sound Vibration* **306** 968–72
[12] Mickens R E 1996 *Oscillations in planar dynamic systems* Series on Advances in Mathematics for Applied Sciences **37** (River Edge, NJ: World Scientific Publishing Co. Inc.) 319 pp
[13] Mickens R E 2010 *Truly Nonlinear Oscillations. Harmonic balance, Parameter Expansions, Iteration, and Averaging Methods* (Hackensack, NJ: World Scientific Publishing Co. Pte. Ltd.) 238 pp
[14] Mickens R E and Wilkerson D 2009 Exact and approximate values of the period for a "truly nonlinear" oscillator: $\ddot{x} + x + x^{1/3} = 0$. *Adv. Appl. Math. Mech.* **1** 383–390
[15] Momeni M, Jamshidi N, Barari A and Ganji D D 2011 Application of He's energy balance method to Duffing-harmonic oscillators *Int. J. Comput. Math.* **88** 135–144.
[16] Müller P C 2008 Comments on the method of harmonic balance for nonlinear conservative single-degree-of-freedom systems *Conference Proceedings, 6th EUROMECH Nonlinear Dynamics Conference (ENOC 2008)* 4 pp.
[17] Rudin W 1987 *Real and complex analysis* Third ed. (New York: McGraw-Hill Book Co.) 416 pp
[18] Stokes A 1972 On the approximation of nonlinear oscillations *J. Differ Equations* **12** 535–58
[19] Urabe M 1964 Relations between periods and amplitudes of periodic solutions of $\ddot{x} + g(x) = 0$ *Funkcial. Ekvac.* **6** 63–88
[20] Urabe M 1965 Galerkin's procedure for nonlinear periodic systems *Arch. Rational Mech. Anal.* **20** 120–52

Modeling cancer evolution: evolutionary escape under immune system control

Andrei Korobeinikov[a,b], **Konstantin E Starkov**[c] **and Paul A Valle**[d]

[a]Centre de Recerca Matematica, Campus de Bellaterra, Edifici C, 08193 Bellaterra, Barcelona, Spain
[b]Departament de Matemàtiques, Universitat Autònoma de Barcelona, Campus de Bellaterra, Edifici C, 08193 Barcelona, Spain
[c]Instituto Politécnico Nacional, CITEDI-IPN, Av. del IPN No. 1310, Mesa de Otay, 22510 Tijuana, B.C., México
[d]Instituto Tecnológico de Tijuana, Blvd. Limón Padilla s/n, Mesa de Otay, Tijuana 22454, B.C., México

E-mail: [a]`akorobeinikov@crm.cat`, [c]`konst@citedi.mx`, [d]`paul.valle@tectijuana.edu.mx`

Abstract. It can be expected that adequate immune response should be able to annihilate cancer at a very early stage of its appearance. However, in some cases cancer is able to persist avoiding immune response. One can conject that cancer is able to avoid immune response control due to a succession of mutations leading to the development of immune-resistant cells. In order to illustrate this possibility, in this paper we present a $2n-$dimensional mathematical model that describes interaction of n subtypes of tumor cells with corresponding genotype-specific immune response. The model postulates that there is a probability for tumor cells of each of n subtype to produce offsprings of other types. Each of the subtypes activates the genotype-specific immune response with a possibility of suppressing cancer cells of other genotypes (the cross-immunity). Numerical simulations show that if cancer cells are able to mutate comparatively fast and if immune response is not strong enough, then, despite immune system pressure, cancer is able to persist.

1. Introduction

Cancer is a general term used to define a group of diseases, which can affect almost any part of the body and is characterized by the uncontrolled growth of abnormal cells that have lost the ability to regulate their lifespan. This phenomenon allows these cells to expand beyond their usual boundaries and invade surrounding tissues and organs, creating new cancer sources, known as malignant tumors. The process in which tumor cells spread throughout the body is called metastasis and is the main reason of cancer patients death [9].

A remarkable feature of cancer is its ability to persist in the presence of immune response, which could be expected to be able to annihilate the abnormal cells at a very early stage of their appearance. It can be conjectured that the ability of cancer to persist under pressure of immune response is related to the ability of cancer cells to mutate. Cancer cells mutate in the cells division process. Eventually, after a succession of mutations, an immune-resistant subtype of cancer cells arises with a specific genotype that allows it and its descendants to evade immune system attacks [2, 10].

In order to illustrate this possibility, in this paper we construct a mathematical model of cancer growth under immune response pressure. In order to develop this mathematical model, we start with the following simple model of cancer-immune response interaction:

$$\frac{dx}{dt} = ax(1-bx) - cxy, \tag{1}$$

$$\frac{dy}{dt} = g\frac{x}{x+s} - \mu y - pxy. \tag{2}$$

In this model, equation (1) describes the growth of tumor cells population of size $x(t)$ while equation (2) models the dynamics of immune effector cells population of size $y(t)$.

The model postulates that in the absence of immune response tumor grows according to the logistic law with a per capita reproduction rate a and the maximum carrying capacity $1/b$. The immune effector cells, of population size $y(t)$, kill cancer cells with a bilinear rate cxy, where c is the immune response strength. The activation of immune response is induced by cancer cells with rate $gx/(x+s)$. The effector cells are removed (are dying) with rate μy. They also can be inactivated due to the interaction with tumor cells (due to exhaustion) with bilinear rate pxy.

Models of this type are fairly common in the literature on cancer dynamics; for example, see [1, 3, 7, 8] and bibliography therein.

The dynamics of system (1), (2) is located in the nonnegative quadrant of a 2-dimensional real space,

$$\mathbf{R}^2_{\geq 0} = \{x \geq 0, y \geq 0\}.$$

It is evident that system (1), (2) has a tumor-free equilibrium state at $(x^*, y^*) = (0, 0)$, and it is easy to verify by the Lyapunov's Indirect Method (see [13]) that this equilibrium is always unstable (a saddle point). This implies that for this particular model the total eradication of the tumor cells population for finite time is impossible. (It can be only considered for infinite time $t \to \infty$.) This is due to the fact that the phase space of the system is continuous with respect to variables $x(t)$ and $y(t)$. However, in the real life there always is a positive finite critical value below which tumor survival is impossible. One can consider this value to be one cancer cell. Then, if there is a solution such that $x(t)$ goes below this threshold (e.g., $x = 1$), one should conclude the complete elimination of the cancer cells population.

2. Mathematical model of cancer evolution

To develop a model of cancer evolution, we make the following assumptions:

(i) The simultaneous existence of multiple cancer cell genotypes in a given population. In this paper we assume the existence of n genotypes.

(ii) A probability for a cancer cell to mutate in the process of cell division. A mutation of a type into another occurs with a certain pre-defined probability. The probability of a mutation of type i to type j (where $i = 1, ..., n$) is lower the larger the difference $|i - j|$ is.

(iii) Specificity of immune response to a cancer cell type. This means that cancer cells of type i activate immune effector cells of type i as well. The activation of immune effector cells is triggered by the presence of cancer cells of a specific genotype and is represented in equation (2) by a term of the Michaelis-Menten kinetics type.

(iv) Cross-immunity. This implies, that immune response activated by cancer cells of type i can also kill cancer cells of other types, but with a lower efficacy. The efficacy of immune effector cells of type i towards cancer cells of type j is inversely proportional to the absolute value $|i - j|$.

Incorporating this set of assumptions into model (1), (2) we obtain the following mathematical model of cancer evolution:

$$\frac{dx_i}{dt} = \left[a_i\left(1 - \sum_{j=1}^{n} A_{ij}\right)x_i + \sum_{j=1}^{n} a_j A_{ji} x_j\right]\left(1 - \sum_{j=1}^{n} b_j x_j\right) - \left(\sum_{j=1}^{n} c_{ji} y_j\right) x_i, \quad (3)$$

$$\frac{dy_i}{dt} = g_i \frac{x_i}{x_i + s_i} - \mu_i y_i - \left(\sum_{j=1}^{n} p_{ji} x_j\right) y_i, \quad (4)$$

where $i = 1, ..., n$.

In this model, the probabilities of mutations of cancer cells are defined by the $n \times n$ diagonally dominant matrix A, where A_{ij} is the probability that a cell of type i produces a cell of type j. The string diagonal dominance of the matrix ensures that the probability of mutation of type i cancer cell into a cell of type j decreases as the distance $|i - j|$ increases. The cross-immunity, that is the ability of immune cells activated by type i cancer cells to kill cancer cells of other types, is defined by the $n \times n$ matrix c_{ji}, which is also diagonal dominant. For simplicity, we assume that the matrix of inactivation rate of the immune effector cells due to their interaction with tumor cells (exhaustion) p_{ji} is proportional to matrix c_{ji}.

For the sake of simplicity, in this paper we assume that cancer cells and immune effector cells of all n types are identical in the sense that values of parameters a, A, b, c, g, p, s and μ are the same for all types. (That is, values of matrix A_{ij}, c_{ij} and p_{ij} depend on the distance $|i - j|$ only and are independent of i.)

The dynamics of model (3)–(4) is located in the nonnegative orthant

$$\mathbf{R}^{2n}_{\geq 0} = \{x_i \geq 0, y_i \geq 0\}.$$

The authors would like to note that for this model the values of the diagonal elements A_{ii} are irrelevant as they are mutually annihilated in equation (3), and are defined for consistency only. One can choose any nonnegative values for these elements, as long as condition $\sum_{j=1}^{n} A_{ij} \leq 1$ is fulfilled.

3. Localization of compact invariant sets

The Localization of Compact Invariant Sets (LCIS) is a method used to analyze global dynamics of mathematical models defined by ordinary differential equations. This approach was recently successfully applied to tumor growth models [6, 11, 12]. Please see [4, 5] and Appendix **A** for a short introduction of this method.

Let us suppose that $\sum_{j=1}^{n} A_{ij} \leq 1$ holds. Then, by the structure of equation (3), one can see that the minimum and the maximum values of each of n tumor subpopulations in the absence of immune response are given by the set

$$K\left(h_{i+n}\right) = \left\{0 \leq x_i \leq x_{i\max} = \frac{1}{b_i}\right\},$$

where $i = 1, ..., n$. We can use this result to estimate ultimate bounds for solutions to equations (4) by applying the LCIS method. Let us now take n localizing functions

$$h_i = y_i, \qquad i = 1, ..., n.$$

Computing its Lie derivative, we find the set

$$S\left(h_i\right) = \left\{\mu_i y_i = \frac{g_i x_i}{x_i + s_i} - \left(\sum_{j=1}^{n} p_{ji} x_j\right) y_i\right\}.$$

such that $S\left(h_i\right) = \left\{L_f h_i = 0\right\}$ holds. The Iterative Theorem yields

$$S\left(h_i\right) \cap K\left(h_{i+n}\right) \cap \mathbf{R}_{\geq 0}^{2n} \subset \left\{y_i \leq \frac{g_i x_{i\max}}{\mu_i\left(x_{i\max} + s_i\right)}\right\},$$

and, therefore, we come to localization set

$$K\left(h_i\right) = \left\{0 \leq y_i \leq \frac{g_i}{\mu_i\left(1 + b_i s_i\right)}\right\}, \qquad i = 1, ..., n.$$

We just proved the following proposition:

Proposition. *All compact invariant sets of system (3), (4) are located inside polytope*

$$K_{BPID} = K\left(h_i\right) \cap K\left(h_{i+n}\right),$$

where

$$K\left(h_{i+n}\right) = \left\{0 \leq x_i \leq x_{i\max} = \frac{1}{b_i}\right\},$$

$$K\left(h_i\right) = \left\{0 \leq y_i \leq y_{i\max} = \frac{g_i}{\mu_i\left(1 + b_i s_i\right)}\right\},$$

and $i = 1, ..., n$.

Proposition implies that for all $t > 0$ all solution to system (3), (4) initiated in the nonnegative orthant remain in this orthant and are bounded. That is, the model is well-defined.

IOP Conf. Series: Journal of Physics: Conf. Series **811** (2017) 012004 doi:10.1088/1742-6596/811/1/012004

4. Numerical simulations

In order to illustrate the idea that mutations and evolution of cancer cells can be accountable for the persistence of cancer despite immune response pressure, we run simulations for system (3), (4) for $n = 20$. Descriptions and values of the model parameters used in these numerical simulations are summarized in Table 1. Please note that the values in Table 1 ensure annihilation of cancer by immune response when there is no possibility of mutation (that is, when all $A_{ij} \equiv 0$).

Table 1. Descriptions, values and units of model parameters.

Parameter	Description	Value and units
a	Cancer cells proliferation rate	4.31×10^{-1} day^{-1}
A	Probability of production a cancer cell of type j by a cell of type i	$--$
b	b^{-1} is tumor carrying capacity	1.02×10^{-14} $cells^{-1}$
c	Killing rate of cancer cells by immune effector cells	3.41×10^{-10} $(cell \times day)^{-1}$
g	Minimal rate of immune cells activation by cancer cells	1.2×10^{4} $cell \times day^{-1}$
s	g/s is maximal rate of immune cells activation by cancer cells	1 $cell$
μ	Death rate of immune cells	5×10^{-4} day^{-1}
p	Effector cells inactivation rate due to exhaustion	2×10^{-11} $(cell \times day)^{-1}$

Figure 1 shows the dynamics of 20 cancer cell subtypes and Figure 2 shows the corresponding cancer-specific immune responses. It can be seen in these Figures that, after a short low-level outbreak of a newly-emerging cancer cell type, specific immune response is activated and eliminates this particular type. However, the activation of genotype-specific immune response requires some time, and during this period, however short it is, new subtypes of cancer cell appear due to mutations. Thus, while immune response is able to control each of the cancer cell subtypes, due to ability to mutate cancer still persists in a host. Figure 3 shows the dynamics of total tumor burden $\sum_{i=1}^{n} x_i(t)$: it can be seen that after a short transition period the tumor size remains constant.

5. Conclusions

Model (3), (4) demonstrates, that, due to its ability to mutate and evolve, cancer is able to persist within a host even when each cancer subtype is eventually annihilated by the immune response. Our simulations show that while each time the immune system controls a tumor subtype, another subtype emerges, and that the total tumor cell levels remain above the threshold of clinical detectability, see [7]. One can expect that, through a process of these continuous mutations, a certain genotype will eventually appear that

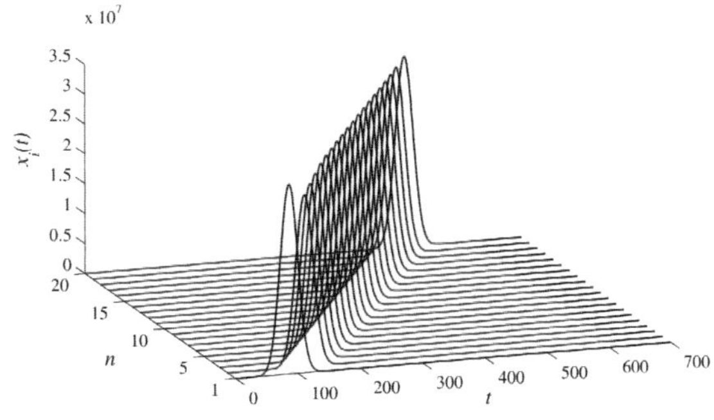

Figure 1. Dynamics of 20 subtypes of cancer cells under a sufficiently strong immune response.

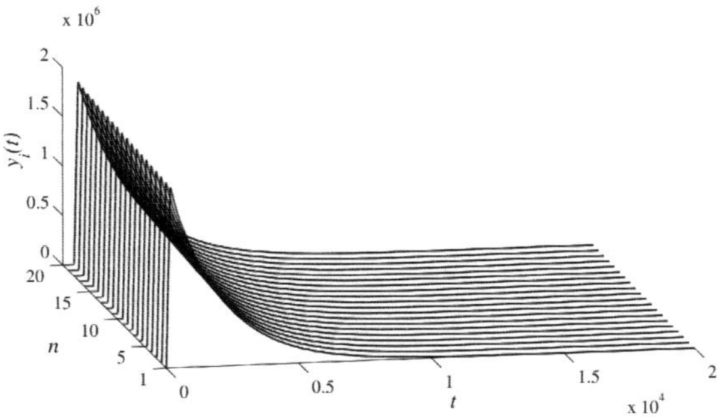

Figure 2. Cancer-specific immune response. Please note that in this figure the time required by a response to settle down to zero level is comparatively large. This reflects the fact that memory cells are present long after disappearance of a pathogen or a cancer cell subtype that activated them.

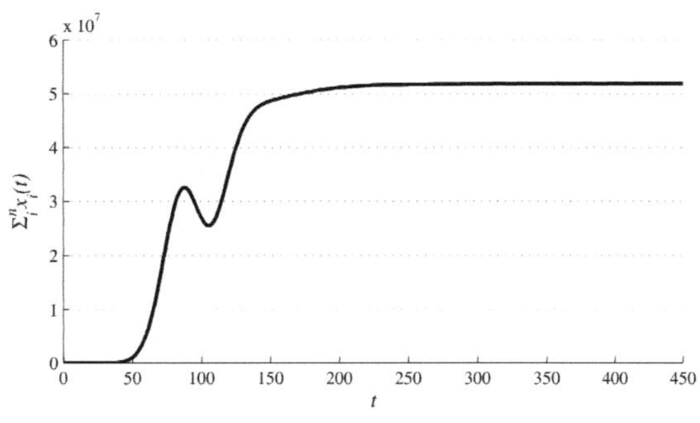

Figure 3. The dynamics of the total tumor burden $\sum_{i=1}^{n} x_i(t)$.

will be able to successfully resist immune control, prevail over other types and grow to the maximum carrying capacity. Then this genotype will be able to induce the angiogenesis process which could lead to the metastasis of tumor cells with all characteristics needed to evade host's immune response.

In these simulations we assumed $n = 20$. Of course, in real life the number of cancer genotypes is higher by several orders of magnitude. However, $n = 20$ is sufficient to illustrate the idea of cancer persistence due to evolutionary scape. Furthermore, for illustrative purpose and to stress the idea that cancer can persists under immune response pressure even without increasing its Darwinian fitness, we assume that all cancer genotypes have the same properties. Of course, in real life situation mutations lead to an increasing diversity of phenotypical traits. Due to natural selection driven by immune response, this diversity leads to increasing Darwinian fitness of cancer cells. In this particular case, the increasing Darwinian fitness means increasing resistance to immune response, till eventually immune-resistant genotype appears.

Acknowledgments
Andrei Korobeinikov is supported by the CERCA Programme of the Generalitat de Catalunya, by the Ministry of Economics of Spain via Ramón y Cajal Fellowship RYC-2011-08061 and grant MTM2015-71509-C2-1-R, by AGAUR of the Generalitat de Catalunya via grant 2014SGR-1307 and by CONACYT (México) via grant 219614. Konstantin Starkov and Paul Valle are supported by the CONACYT, México, via project number 219614. Paul Valle is also supported by the CONACYT scholarship 350156.

Appendix A. Localization of compact invariant sets method

The Localization of Compact Invariant Sets method is used to determine a domain in the state space where all compact invariant sets are located. As examples of compact invariant sets, one may recall equilibrium points, periodic orbits, homoclinic and heteroclinic orbits, invariant tori and chaotic attractors. The relevance of this analysis is due to the fact that it is useful to study the long-time dynamics of a system.

Let us consider an autonomous nonlinear system of the form:

$$\dot{x} = f(x), \tag{A.1}$$

where f is a C^∞−differentiable vector function and $x \in \mathbf{R}^n$ is the state vector. Let $h(x) : \mathbf{R}^n \to \mathbf{R}$ be a C^∞−differentiable function, by $h|_B$ we denote the restriction of h on a set $B \subset \mathbf{R}^n$. The function h used in this statement is called *localizing* function, and we assume that h is not a first integral of (A.1). By $S(h)$ we denote the set $\{x \in \mathbf{R}^n \mid L_f h(x) = 0\}$, where $L_f h(x) = (\partial h/\partial x) f(x)$ represents the Lie derivative of (A.1). Let us define $h_{\text{inf}} := \inf\{h(x) \mid x \in S(h)\}$ and $h_{\text{sup}} := \sup\{h(x) \mid x \in S(h)\}$. Then the general theorem concerning the localization of all compact invariant sets of a dynamical system establishes the following

Theorem A.1. See [4, 5]. *Each compact invariant set* Γ *of (A.1) is contained in the localization set* $K(h) = \{h_{\text{inf}} \leq h(x) \leq h_{\text{sup}}\}$.

If we consider the location of all compact invariant sets inside domain $U \subset \mathbf{R}^n$, we have set $K(h) \cap U$, with $K(h)$ defined by Theorem A.1. It is evident that if all compact invariant sets are located in the sets Q_1 and Q_2, with $Q_1, Q_2 \subset \mathbf{R}^n$, then they are located in the set $Q_1 \cap Q_2$ as well.

Suppose that we are interested in the localization of all compact invariant sets located in some subset Q of the state space $\mathbf{R}^n$. We formulate the following Proposition:

Proposition A.2. (See [4, 5].) *If $Q \cap S(h) = \emptyset$, then the system (A.1) has no compact invariant sets located in Q.*

A refinement of the localization set $K(h)$ is realized with help of the Iterative Theorem stated as follows:

Theorem A.3. (See [4, 5].) *Let $h_m(x), m = 0, 1, 2, \dots$ be a sequence of $C^\infty-$differentiable functions. Sets*

$$K_0 = K(h_0), \quad K_m = K_{m-1} \cap K_{m-1,m}, \quad m > 0,$$

with

$$\begin{aligned} K_{m-1,m} &= \{x : h_{m,inf} \le h_m(x) \le h_{m,sup}\}, \\ h_{m,sup} &= \sup_{S(h_m)\cap K_{m-1}} h_m(x), \\ h_{m,inf} &= \inf_{S(h_m)\cap K_{m-1}} h_m(x), \end{aligned}$$

contain any compact invariant set of the system (A.1) and

$$K_0 \supseteq K_1 \supseteq \dots \supseteq K_m \supseteq \dots .$$

References

[1] Arciero J, Jackson T and Kirschner D 2004 A mathematical model of tumor-immune evasion and siRNA treatment. *Discrete Cont. Dyn.-B* **4** 39–58.
[2] Kim R, Emi M and Tanabe K 2007 Cancer immunoediting from immune surveillance to immune escape. *Immunology* **121** 1–4.
[3] Kirschner D and Panetta JC 1998 Modeling immunotherapy of the tumor–immune interaction. *J. Math. Biol.* **37** 235–252.
[4] Krishchenko AP 2005 Localization of invariant compact sets of dynamical systems . *Diff. Equat+* **41** 1669–1676.
[5] Krishchenko AP and Starkov KE 2006 Localization of compact invariant sets of the Lorenz system. *Phys Lett A* **353** 383–388.
[6] Krishchenko AP and Starkov KE 2016 On the global dynamics of a chronic myelogenous leukemia model. *Commun. Nonlinear Sci. Numer. Simul.* **33** 174-83.
[7] de Pillis L, Gu W and Radunskaya A 2006 Mixed immunotherapy and chemotherapy of tumors: modeling, applications and biological interpretations. *J. Theor. Biol.* **238** 841–862.
[8] de Pillis L, Gu W, Fister K, Head T, Maples K, Murugan A, et al. 2007 Chemotherapy for tumors: An analysis of the dynamics and a study of quadratic and linear optimal controls. *Math. Biosci.* **209** 292–315.
[9] Ruddon RW 2007 *Cancer biology*(Oxford University Press)
[10] Schreiber RD, Old LJ and Smyth MJ 2011 Cancer immunoediting: integrating immunity's roles in cancer suppression and promotion. *Science* **331** 1565–1570.
[11] Starkov KE and Krishchenko AP 2014 On the global dynamics of one cancer tumour growth model. *Commun. Nonlinear Sci.* **19** 1486–1495.
[12] Valle PA, Starkov KE and Coria LN 2016 Global stability and tumor clearance conditions for a cancer chemotherapy system. *Commun. Nonlinear. Sci. Numer. Simul.* **40** C, 206–215.
[13] Khalil HK 2002 *Nonlinear Systems* (Prentice Hall).

IOP Conf. Series: Journal of Physics: Conf. Series **811** (2017) 012005 doi:10.1088/1742-6596/811/1/012005

A continuous phenotype space model of cancer evolution

David Masip[1,2,a], **Andrei Korobeinikiov**[1,2,b]

[1]Centre de Recerca Matemàtica, Campus de Bellaterra, Edifici C, 08193 Bellaterra, Barcelona, Spain

[2] Departament de Matemàtiques, Universitat Autònoma de Barcelona, Campus de Bellaterra, Edifici C, 08193 Bellaterra, Barcelona, Spain

E-mail: [a]david.masipb@e-campus.uab.cat, [b]akorobeinikov@crm.cat.

Abstract. It was suggested that the ability of cancer to avoid immune response pressure (that should be expected to be capable to annihilate cancer at its early stage) can be attributed to the ability of the cancer cells to evolve. The goal of this notice is to illustrate this possibility by the means of mathematical modelling. In this notice, we construct a simple mechanistic model of cancer evolution, which is based upon a classical model of cancer-immune response interaction. Numerical simulations confirm the hypothesis that if cancer mutates fast enough and if immune response is not sufficiently strong, then cancer is able to avoid immune response pressure by evolution.

1. Introduction

The very general term "cancer" usually refer a set of diseases caused by an atypical growth of a certain kind of cells. This cells are typically replicate considerably faster than normal cells do, and are able to mutate quickly. Such high mutability enables cancer cells under some circumstances to mislead the immune system and avoid immune response pressure (that normally should be expected to be able to clean the cancer up).

The objective of this notice is to explore the possibility of cancer evolutionary escape off the immune response pressure. With this objective in mind, we construct a simple model of cancer evolution, which is based on well-known models of cancer-immune response interaction [3, 5, 4, 8]. To develop a model of cancer evolution we are using an approach that was developed in [1, 2, 6]. In these papers that were dealing with within-host viral evolution, simultaneous existence of a multitude of pathogen genotypes was postulates, and these multiple genotype was arranged into a continuous variant space. In this space distance between variants can be defined in some way, and, hence, random mutations can be modeled. In [1, 2, 6] random mutations were modeled by diffusion in the continuous variant space. However, taking into consideration the particularities of the model, in this notice we find that it would be more appropriate to use an integral operator to model the mutations. Applying this general approach to model in [3, 5, 4, 8], we derive a PDE model of cancer evolution. We perform a non-dimensionalization of this model, and, finally, we numerically integrate the differential equations of the model for given initial conditions. Numerical simulations confirm the hypothesis that the ability of cancer to evolve can be responsible for its persistence under immune response pressure.

2. Model

To derive a model of cancer evolution, we use a simple classical model that describes the interaction between cancer cells and immune response (see [3, 5, 8]):

$$\frac{dx}{dt} = ax\left(1 - \frac{x}{k}\right) - \beta xy, \tag{1}$$

$$\frac{dy}{dt} = c\frac{x}{\gamma + x}(1 + ey) - my - b\beta xy. \tag{2}$$

In this model, $x(t)$ and $y(t)$ are concentrations of cancer cells and immune response cells, respectively, at time t. The other parameters of the model are positive constants. This model postulates a logistic growth of cancer cells, with per capita reproduction rate a and carrying capacity k. The term βxy accounts for killing of the cancer cells by immune effector cells, and hence parameter β is the efficiency of the immune response against this type of cancer cells. The term $c\frac{x(t)}{\gamma+x(t)}(1 + ey(t))$ represents proliferation of the immune response cells. Specifically, factor $\frac{x(t)}{\gamma+x(t)}$ accounts for the activation of immune response by cancer cells. For low levels of $x(t)$, the activation of the immune response system is proportional to $x(t)$. As $x(t)$ grows, the rate of activation grows asymptotically to a certain constant level. Apart from the activation of immune response, term $c\frac{x(t)}{\gamma+x(t)}(1 + ey(t))$ also takes account of the proliferation of immune response cells by cloning. Term my represents the natural death of the immune effector cells; thus, $1/m$ is an average life time of the immune cells. Term $b\beta xy$ takes into account the death of immune cells by exhaustion. This implies that an immune cell can kill only a limited number of cancer cells, and then, after a certain number of killings, the cell gets exhausted and dies by apoptosis.

In order to construct a model of evolution on the basis of this model, we assume the existence of multiple variants of cancer cells and postulate that these are distributed in a continuous variant space. In this notice we assume that the variant space is one-dimensional and is parameterized by a parameter s. Similarly, immune response cells, which emerge in response to a specific variant of cancer cells, are also represented by their density in the same continuous space of variants. We denote by $x(t, s)$ the density distribution of cancer cells by variants s at time t, and, likewise, by $y(t, s)$ the density distribution of the immune response cells by variants s at time t.

We assume that a cancer cell produces new cells in the process of cell division. In this process, the majority of the daughter cells are of the same variant. However, with some probability, due to mutations a cell is also able to produce daughter cells of other variants. Although this process is probabilistic, we are using a deterministic model, as ultimately we are only interested in evolution of cancer cells.

Let us denote by $a(s)$ the per capita reproduction rate of variant s and by $\xi(r, s)$ the probability that a variant s produces a cell of variant r. As above, k is the carrying capacity. Then, the simplest generalization of the logistic growth is the following growth rate of the variant s at time t:

$$\left(1 - \frac{\int_0^\infty x(t, r)dr}{k}\right)\int_0^\infty a(r)\xi(s, r)x(t, r)dr. \tag{3}$$

Here the value of $\int_0^\infty a(r)\xi(s, r)x(t, r)dr$ is the sum of all mutations over all possible variants into variant s. The kernel $\xi(s, r)$ satisfies the equalities

$$\int_0^\infty \xi(s, r)ds = \int_0^\infty \xi(s, r)dr = 1.$$

In this notice, we use a gaussian distribution with center r and small variance to model $\xi(s,r)$. However, other kernels are also possible. Furthermore, while the reproduction rate is defined in (3) as a function $a(s)$, further in this paper we will set $a(s) = a = const$, as there is no real-life evidence that the reproduction rate significantly changes with variant.

Factor $\left(1 - \frac{\int_0^\infty x(t,r)dr}{k}\right)$ models the limitation of the growth due to the limited carrying capacity of the tumor. This term implies that all variants consume the same amount of resources. Here $\int_0^\infty x(t,r)dr$ is the total amount of cancer cells at time t.

We assume that a cancer variant actives variant-specific immune response cells, and hence the immune cells activated in this process are labeled with the same s as the cancer cells that activate these particular cells. However, at the same time we assume that immune response cells can with some probability kill cancer cells of variants different from the one that activated them. That is, we take into consideration the so-called cross-immunity. Let us denote by $\beta(s)$ the efficacy of the immune response cells of variant s against the cancer variant s. We define $\eta(s,r)$ as the relative efficacy of the immune response cells of variant s when killing cancer cells of variant r. Then, the simplest generalization of the bilinear term βxy in (1) is

$$x(t,s) \cdot \int_0^\infty \beta(r)\eta(r,s)y(t,r)dr. \tag{4}$$

Please note that in this case $\eta(r,s)$ is not a probability distribution function (that is, it is not a kernel), but a function that satisfies $\eta(s,s) = 1$ whereas $\eta(r,s) < 1$ holds for all $r \neq s$. To model this function, we use an unnormalized gaussian function $\eta(r,s) = e^{(r-s)^2/(2\sigma^2)}$. The value σ can be viewed as a typical width of the function, which is assumed small and decreasing as $|r - s|$ grows.

Using equations (3) and (4) we can derive a differential equation for cancer cells growth as following:

$$\begin{aligned} \frac{\partial x(t,s)}{\partial t} &= \left(1 - \frac{\int_0^\infty x(t,r)dr}{k}\right)\int_0^\infty a(r)\xi(s,r)x(t,r)dr- \\ &- \int_0^\infty \beta(r)\eta(r,s)y(t,r)x(t,s)dr. \end{aligned} \tag{5}$$

In order to make the system consistent we have now to define an equation for $y(t,s)$. To derive this equation, we are going to consider activation, reproduction, natural death and the exhaustion of immune cells. As in model (2), we model the growth of immune cells of variant s at time t by term

$$c\frac{x(t,s)}{\gamma + x(t,s)}(1 + ey(t,s)). \tag{6}$$

For the death of the immune cells, we consider both natural death, with rate $my(t,s)$, and death due to exhaustion, which in this case is represented by term

$$b\int_0^\infty \beta(s)\eta(s,r)x(t,r)y(t,s)dr. \tag{7}$$

That is, this term implies that killing rate of an immune effector cell of type s is, on average, $\int_0^\infty \beta(s)\eta(s,r)x(t,r)dr$ cancer cells. Please note that for a given variant s, this term summarizes all the cancer cells that get killed by this particular immune cell variant, whereas in equation (5), in contrast, we summarize, for a given cancer variant s, all the immune cells that kill cancer cells of this variant. Finally, we assume here that immune effector cells of all variants kill an

IOP Conf. Series: Journal of Physics: Conf. Series **811** (2017) 012005 doi:10.1088/1742-6596/811/1/012005

average of $1/b$ cancer cells before get exhausted, which is the reason why the factor b appears in the equation. With these assumptions, using equations (6) and (7), we finally come to equation

$$\frac{\partial y(t,s)}{\partial t} = c\frac{x(t,s)}{\gamma + x(t,s)}(1 + ey(t,s)) - my(t,s) - b\int_0^\infty \beta(s)\eta(s,r)x(t,r)y(t,s)dr \tag{8}$$

It may be noteworthy that this model is equivalent to a discrete variant space model reported in [7].

3. Non-dimensionalization of the equations

Let us define a set of non-dimensional variables $\bar{x}$, $\bar{y}$, $\bar{t}$ and $\bar{s}$, and a set of constants (scales) X, Y, T and S, such that

$$t = \bar{t}T \Rightarrow \bar{t} = \frac{t}{T}, \tag{9}$$

$$s = \bar{s}S \Rightarrow \bar{s} = \frac{s}{S}, \tag{10}$$

$$x(t,s) = \bar{x}(\bar{t},\bar{s})X \Rightarrow \bar{x}(\bar{t},\bar{s}) = \frac{x(t,s)}{X}, \tag{11}$$

$$y(t,s) = \bar{y}(\bar{t},\bar{s})Y \Rightarrow \bar{y}(\bar{t},\bar{s}) = \frac{y(t,s)}{Y}. \tag{12}$$

Constant S does not appear explicitally in any of the equations (and hence, in fact, variable s is non-dimensional), and hence we set $S = 1$. Then, after substitution of (9)–(12) in (5), (8), the equations of our model are

$$\begin{aligned}\frac{X}{T}\frac{\partial \bar{x}(\bar{t},\bar{s})}{\partial \bar{t}} &= \left(1 - \frac{X\int_0^\infty \bar{x}(\bar{t},\bar{r})d\bar{r}}{k}\right)\int_0^\infty aX\xi(\bar{s},\bar{r})\bar{x}(\bar{t},\bar{r})d\bar{r} \\ &\quad -\bar{x}(\bar{t},\bar{s})XY\int_0^\infty \beta(\bar{r})\eta(\bar{r},\bar{s})\bar{y}(\bar{t},\bar{r})d\bar{r},\end{aligned} \tag{13}$$

$$\begin{aligned}\frac{Y}{T}\frac{\partial \bar{y}(\bar{t},\bar{s})}{\partial \bar{t}} &= c\frac{X\bar{x}(\bar{t},\bar{s})}{\gamma + X\bar{x}(\bar{t},\bar{s})}(1 + eY\bar{y}(\bar{t},\bar{s})) - mY\bar{y}(\bar{t},\bar{s}) \\ &\quad -bXY\beta(\bar{s})\bar{y}(\bar{t},\bar{s})\int_0^\infty \eta(\bar{s},\bar{r})\bar{x}(\bar{t},\bar{r})d\bar{r}.\end{aligned} \tag{14}$$

Let us assume that

$$X = k, \qquad T = \frac{1}{a}, \qquad Y = cT = \frac{c}{a} \tag{15}$$

and define non-dimensional parameters

$$\bar{\beta}(\bar{s}) = \frac{c}{a^2}\cdot\beta(s), \qquad \bar{m} = \frac{m}{a}, \qquad \bar{\gamma} = \frac{\gamma}{k}, \qquad \bar{e} = e\frac{c}{a}, \qquad \bar{b} = b\frac{ka}{c}. \tag{16}$$

Substituting (15) and (16) into system (13), (14), we obtain, after obvious algebraic simplification, non-dimensional system

$$\frac{\partial \bar{x}(\bar{t},\bar{s})}{\partial \bar{t}} = \left(1 - \int_0^\infty \bar{x}(\bar{t},\bar{r})d\bar{r}\right)\int_0^\infty \xi(\bar{s},\bar{r})\bar{x}(\bar{t},\bar{r})d\bar{r} - \bar{x}(\bar{t},\bar{s})\int_0^\infty \bar{\beta}(\bar{r})\eta(\bar{r},\bar{s})\bar{y}(\bar{t},\bar{r})d\bar{r}, \tag{17}$$

$$\frac{\partial \bar{y}(\bar{t},\bar{s})}{\partial \bar{t}} = \frac{\bar{x}(\bar{t},\bar{s})}{\bar{\gamma} + \bar{x}(\bar{t},\bar{s})}(1 + \bar{e}\bar{y}(\bar{t},\bar{s})) - \bar{m}\bar{y}(\bar{t},\bar{s}) - \bar{b}\bar{\beta}(\bar{s})\bar{y}(\bar{t},\bar{s})\int_0^\infty \eta(\bar{s},\bar{r})\bar{x}(\bar{t},\bar{r})d\bar{r}. \tag{18}$$

4. Numerical simulations

To carry out a numerical simulations, we have to define function $\beta(s)$. We assume, in order to compare our results with the results in [7], that function $\beta(s)$ is constant, $\beta(s) = \beta$. This implies that all cancer variants are the same. Furthermore, to integrate the differential equations one also needs to define initial and boundary conditions. The initial condition that we use is fairly simple: we assume that initially there are 100 cancer cells of a single variant and no cells of other variants. Specifically, we assume that there are 100 cells at point $s = 0.5$, and that at all other points the cancer cell concentration is equal to zero.

We carried out numerical simulations of the model using parameters from [5] and [8]. The values of parameters are summorized in Table 1. The parameters σ_{mut} and σ_{immu} are the variance of the gaussian distributions that model mutation rate $\xi(s, r)$ and the width of cross-immunity $\eta(s, r)$.

Table 1. Values of the parameters used in simulations (from [5, 8]).

Meaning	Notation	Value	Units
Cancer cells proliferation rate	a	$4.3 \cdot 10^{-3}$	day^{-1}
Carrying capacity of the tumor	k	10^{14}	cells
Immune cells natural death rate	m	0.0412	day^{-1}
Immune cells killing rate	β	$3.4 \cdot 10^{-10}$	day^{-1}·cells^{-1}
Steepness coefficient of the immune cell recruitment curve	γ	$2 \cdot 10^{7}$	cells
Maximum immune cell activation rate by tumor cells	c	$1.2 \cdot 10^{4}$	cells·day^{-1}
Immune cells proliferation rate	e	$5 \cdot 10^{-4}$	cells^{-1}
$\frac{1}{b}$ is an average number of killing before an immune cell is exhausted	b	0.1	–
Standard deviation of the mutation probability distribution	σ_{mut}	0.02	–
Width of relative efficacy function	σ_{immu}	0.04	–

In the non-dimensional model (17), (18), the values of parameters change as described in Section 3. In particular, in these equation $\bar{x}(\bar{t}, \bar{s})$ represents a fraction of the maximal size of the tumor, k, which is, in our simulation, of 10^{14} cells. Similarly, the value of $\bar{y}(\bar{t}, \bar{s})$ represents a fraction of the maximal immune response, $\frac{c}{a} \approx 2.8 \cdot 10^{6}$ immune cells. Non-dimensional time $\bar{t}$ is a fraction of approximately 232 days. (The non-dimensional time $\bar{t}$ and physical time t are related as $t = \bar{t}/a$, and for $a = 4.3 \cdot 10^{-3}$ in Table 1 $1/a \approx 232$ days.) The values of the parameters in the non-dimensional model are given in Table 2.

Table 2. Values of the parameters in non-dimensional model (17), (18).

Parameter	$\bar{m}$	$\bar{\gamma}$	$\bar{\beta}$	$\bar{e}$	$\bar{b}$	σ_{mut}	σ_{immu}
Value	9,58	$2 \cdot 10^{-7}$	0.22	$1.4 \cdot 10^{3}$	$3.58 \cdot 10^{6}$	0.02	0.04

The results of numerical simulation are shown in Figures 1 and 2. Please note that, although our computations have been run from $t = 0$, the concentration of cancer cells for $t < 0.03$ is so low, compared to the levels reached later, that it cannot be seen in the figures. Therefore we have plotted the solution for $t > 0.03$. These Figures clearly demonstrate the formation of two symmetric pulse-type travelling waves of evolution going into opposite directions. Two traveling waves, rather than a single wave, form in these simulations because the fitness (which for this model is $\beta(s)$) of all cancer variants was assumed to be the same, and, hence, both directions of the variant space are equivalent. These simulations demonstrate that the model can be used for cancer evolution simulations and confirm that evolution enables cancer to avoid the immune response pressure.

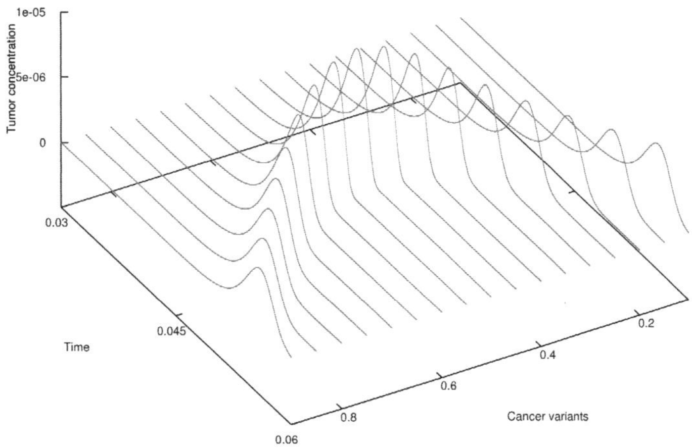

Figure 1. Formation of two symmetric pulse-type traveling waves of evolution going in opposite directions from the initial variant at $s = 0.5$.

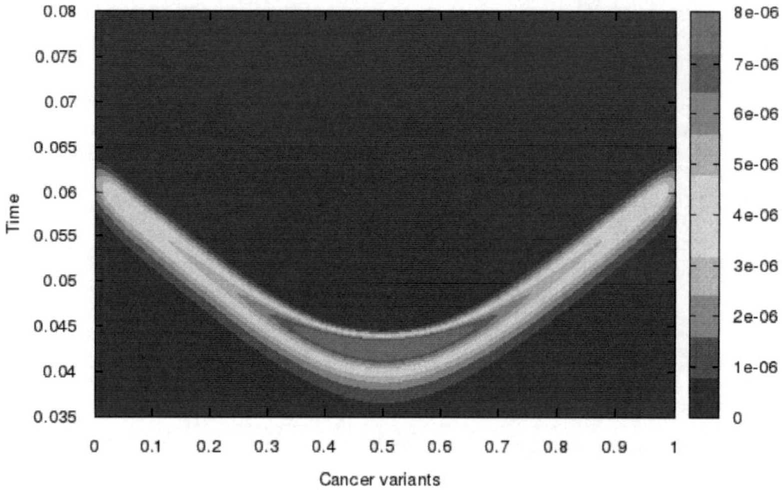

Figure 2. The formation and propagation of two traveling wave of evolution, going in opposite directions in the variant space from the initial variant at $s = 0.5$. In this picture, the concentration of cancer cells is represented by color; see legend on the right-hand side of the plot.

Acknowledgments
Andrei Korobeinikov is supported by the CERCA Programme of the Generalitat de Catalunya, by the Ministry of Economics of Spain via Ramón y Cajal Fellowship RYC-2011-08061 and grant MTM2015-71509-C2-1-R, by AGAUR of the Generalitat de Catalunya via grant 2014SGR-1307 and by CONACYT (México) via grant 219614.

David Masip is a undergraduate intern at the CRM.

References
[1] Archibasov AA, Korobeinikov A and Sobolev VA 2015 Asymptotic expansions of solutions for a singularly perturbed model of viral evolution. *Computational Mathematics and Mathematical Physics* **55** (2), 240–250.
[2] Archibasov AA, Korobeinikov A and Sobolev VA 2016 Passage to the limits in a singularly perturbed partial integro-differential system. *Differential Equations* **52** 1115–1122.
[3] Arciero JC, Jackson TL and Kirschner DE 2004 A mathematical model of tumor-immune evasion and siRNA treatment. *Discrete Cont. Dyn.-B* **4** 39–58.
[4] de Pillis L, Gu W and Radunskata A 2006 Mixed immunotherapy and chemotherapy of tumors: modeling applications and biological interpretations. *J. Theor. Biol.* **238** 841–862.
[5] de Pillis L, Gu W, Fister K, Head T, Maples K, Murugan A, Neal T, and Yoshida K 2007 Chemotherapy for tumors: an analysis of the dynamics and a study of quadratic and linear optimal controls. *Math. Biosci.* **209** 292–315.
[6] A. Korobeinikov and C. Dempsey, A continuous phenotype space model of RNA virus evolution within a host, *Math.Biosci.Eng. 11(4), 919-927, 2014.*
[7] Korobeinikiov A Starkov KE and Valle PA Modeling cancer evolution: evolutionary escape under immune system control. *Journal of Physics Conference Series.* (to appear).
[8] Kuznetsov VA, Makalkin IA, Taylor MA and Perelson AS 1994 Nonlinear Dynamics of Immunogenic Tumors: Parameter Estimation and Global Bifurcation Analysis. *Bull. Math. Biol.* **56** 295–321.

IOP Conf. Series: Journal of Physics: Conf. Series **811** (2017) 012006 doi:10.1088/1742-6596/811/1/012006

Nonlinearization and waves in bounded media: old wine in a new bottle

Michael P Mortell[1] **and Brian R Seymour**[2]

Abstract. We consider problems such as a standing wave in a closed straight tube, a self-sustained oscillation, damped resonance, evolution of resonance and resonance between concentric spheres. These nonlinear problems, and other similar ones, have been solved by a variety of techniques when it is seen that linear theory fails.

The unifying approach given here is to initially set up the appropriate linear difference equation, where the difference is the linear travel time. When the linear travel time is replaced by a corrected nonlinear travel time, the nonlinear difference equation yields the required solution.

[1] Department of Applied Mathematics, University College, Cork, Ireland
[2] Department of Mathematics, University of British Columbia, Vancouver, Canada

E-mail: m.mortell@ucc.ie, seymour@math.ubc.ca

1. Introduction

The series of problems discussed here all involve small amplitude, one-dimensional, nonlinear hyperbolic waves in a medium of finite extent, and thus inherently involve wave reflections from the boundaries. Guided by experiments such as in Saenger & Hudson [1], our fundamental hypothesis is that, for small amplitude disturbances, the interaction of these nonlinear waves is negligible in calculating the main features of the flow, such as the presence of shocks. Then we can frame the hypothesis as: "the motion consists of non-interacting simple waves."

In all examples presented here linear theory fails; for unforced standing waves linear theory fails in the long term as it fails to predict a singularity, and for forced periodic oscillations linear theory fails to give a bounded periodic solution at resonance. The presence of shocks in the flows is central to the observations.

We consider standing waves, a self-sustained oscillation, periodic resonance and its evolution, all in the context of a gas contained in a long straight tube. Similar resonance problems between concentric spheres and cylinders are also considered, making use of nonlinear geometric acoustics approximations. These problems have been solved previously by a variety of techniques. The contribution made here is a simple observation that allows for a simpler method of solution to each problem: "correct the travel time in linear theory by an appropriate *nonlinear travel time*," when allied to "the motion consists of non-interacting simple waves."

The genesis of this idea goes back to Whitham [2] who considered the supersonic flow past a projectile. He introduces "the fundamental hypothesis that linearized theory gives a valid first approximation to the flow *everywhere* provided that in it the approximate characteristics are replaced by the exact ones, or at least a sufficiently good approximation to the exact ones." This is referred to as Whitham's nonlinearization technique, see Whitham [3], Landau [4] and Lighthill [5].

We extend these ideas to reflected waves in a bounded medium by using approximate nonlinear characteristics to calculate the nonlinear travel time, which depends on the signal carried, to replace the linear travel time, i.e., the time for a wave to complete a round trip in the bounded medium. This calculation is carried out on the basis that the motion in the medium is the superposition of two non-interacting simple waves.

These observations allow us to propose a simple and unified approach to the solution of the various problems.

2. Basic equations

The oscillations of a polytropic gas in a closed tube are described in terms of the equations of conservation of mass and momentum relating the velocity $u(x,t)$ and density $\rho(x,t)$. In Eulerian coordinates they are:

$$\frac{\partial \rho}{\partial t} + \frac{\partial}{\partial x}(u\rho) = 0, \quad \frac{\partial u}{\partial t} + u\frac{\partial u}{\partial x} + \rho^{-1}\frac{\partial p}{\partial x} = 0. \tag{1}$$

Pressure and density are measured from their values in a reference state (p_0, ρ_0), so that

$$\frac{p}{p_0} = (\frac{\rho}{\rho_0})^{\gamma} = (1+e)^{\gamma} = 1 + \gamma e + \frac{\gamma(\gamma-1)}{2}e^2 + ... \tag{2}$$

where $e(x,t) = \rho/\rho_0 - 1$ is the condensation, and $c_0 = \sqrt{\frac{\gamma p_0}{\rho_0}}$ the associated sound speed.

Linear Theory

When $|e| \ll 1$ and $|u| \ll c_0$, the corresponding linear equations are

$$\frac{\partial e}{\partial t} + \frac{\partial u}{\partial x} = 0, \text{ and } \frac{\partial u}{\partial t} + c_0^2\frac{\partial e}{\partial x} = 0. \tag{3}$$

Eliminating e, u satisfies the linear wave equation:

$$\frac{\partial^2 u}{\partial t^2} - c_0^2\frac{\partial^2 u}{\partial x^2} = 0. \tag{4}$$

Velocity, pressure and density are nondimensionalized with respect to $(c_0, \rho_0 c_0^2, \rho_0)$, and (u, p, ρ) are considered as functions of length and time $(Lx, Lc_0^{-1}t)$, where L is the tube length. Then equation (4) becomes

$$\frac{\partial^2 u}{\partial t^2} - \frac{\partial^2 u}{\partial x^2} = 0, \ 0 \le x \le 1, \ t > 0. \tag{5}$$

The general solution to equation (5) is

$$u(\alpha, \beta) = f(\alpha) + g(\beta), \ e = f(\alpha) - g(\beta), \tag{6}$$

where f and g are arbitrary functions and

$$\alpha = t - x, \ \beta = t + x - 1, \tag{7}$$

are the linear characteristics. Note that $\alpha = t$ on $x = 0$; $\beta = t$ on $x = 1$.

Nonlinear Theory

The sound speed is $a(e) = (1+e)^{(\gamma+1)/2}$, so that, in dimensionless variables, when $|e| \ll 1$,

$$a(e) = 1 + \frac{\gamma+1}{2}e + 0(e^2). \tag{8}$$

For an $\alpha-$ wave traveling to the right, with no $\beta-$ wave, $e = f(\alpha)$ by (6), and

$$\begin{aligned}\frac{1}{a} &= \frac{dt}{dx}|_\alpha = 1 - \frac{\gamma+1}{2}e + 0(e^2)\\ &= 1 - \frac{\gamma+1}{2}f(\alpha) + ..\end{aligned}$$

Integrating:

$$t = \alpha + x - \frac{\gamma+1}{2}xf(\alpha) + ..$$

with $t = \alpha$ on $x = 0$. Hence the approximate right-traveling nonlinear characteristic, $\alpha =$ const., is given by

$$\alpha = t - x + \frac{\gamma+1}{2}xf(\alpha) + .. \tag{9}$$

Similarly, the approximate left-traveling nonlinear characteristic, $\beta =$ const., is given by

$$\beta = t + (x-1) + \frac{\gamma+1}{2}(x-1)g(\beta) + .. \tag{10}$$

with $\beta = t$ on $x = 1$.

The fundamental assumption in the calculations for the nonlinear characteristics α and β in (9) and (10) is that the α and β waves do not interact to this order of approximation.

3. Standing wave in a closed tube

For a standing wave in a closed tube, the boundary conditions are

$$u(0,t) = 0, \ u(1,t) = 0. \tag{11}$$

In linear theory, using (6) and (7) in the boundary conditions (11) implies that $f(t) = -g(t-1)$ and $g(t) = -f(t-1)$, so g satisfies the linear difference equation

$$g(t) - g(t-2) = 0. \tag{12}$$

Then g, f and u have period 2 in time and maintain their initial form given at $t = 0$. This contradicts the results from Lax [6] who showed from the exact equations that a singularity must arise.

Consider a wave that leaves $x = 1$ at time $t = t_0$, is reflected at $x = 0$ at time $t = t_1$ and arrives back at $x = 1$ at time $t = t_2$. At each reflection the boundary condition is $u = 0$, hence from (6) at each boundary $g = -f$. The travel time down the tube and back is $t_2 - t_0$.

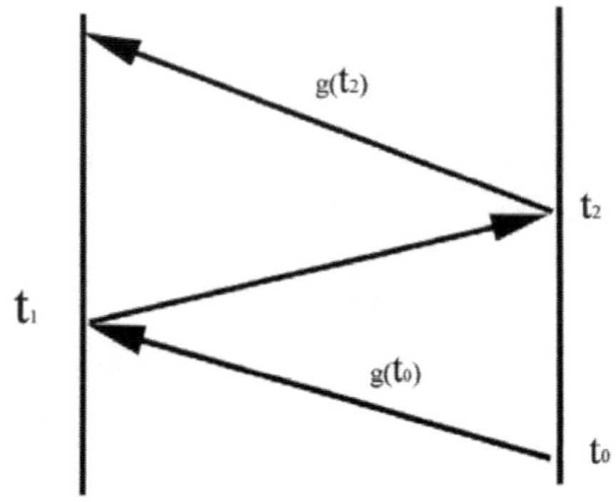

Characteristics between t_0 and t_2.

Nonlinear Travel Time

For the characteristic $\beta = t_0$ leaving $x = 1$, (10) and (11) imply that on $x = 0$

$$t_0 = t_1 - 1 - \frac{\gamma+1}{2} g(t_0) \text{ and } f(t_1) = -g(t_0). \tag{13}$$

For the characteristic $\alpha = t_1$ leaving $x = 0$, (9) and (11) imply that on $x = 1$

$$t_1 = t_2 - 1 + \frac{\gamma+1}{2} f(t_1) \text{ and } f(t_1) = -g(t_2). \tag{14}$$

Hence

$$t_2 = t_0 + 2 + \frac{\gamma+1}{2}[g(t_0) + g(t_2)] \text{ and } g(t_2) = -f(t_1) = g(t_0). \tag{15}$$

Hence the *corrected* nonlinear travel time is

$$t_2 - t_0 = 2 + (\gamma + 1)g(t_0). \tag{16}$$

Substituting this into the linear difference equation (12) gives

$$g(t) - g(t + 2 + (\gamma + 1)g(t)) = 0. \tag{17}$$

This represents a *simple wave* signal distortion on the boundary $x = 1$. A smooth initial function $g(t)$ will break to give a shock, agreeing with Lax [6]; see also Mortell & Varley [7].

4. Damped standing wave in a closed tube

While the end $x = 1$ is again kept fixed, so $u(1, t) = 0$ and $g(t) = -f(t - 1)$, energy is allowed to radiate out through the surface at $x = 0$ via the interface condition

$$e(0, t) = -iu(0, t). \tag{18}$$

This implies that at $x = 0$

$$f(t) = -kg(t - 1), \; k = \frac{i-1}{i+1}, \; 0 \leq i < \infty, \tag{19}$$

where k $(-1 < k < 1)$ is the reflection coefficient. Then (12) becomes the linear difference equation

$$g(t) - kg(t-2) = 0, \tag{20}$$

that implies $|g(t)| \to 0$ as $t \to \infty$.The signal is attenuated geometrically like $|k|^n$.

Nonlinear Travel Time

Using (19) the nonlinear travel time (16) becomes

$$t_2 - t_0 = 2 + (1+k)\frac{(\gamma+1)}{2}g(t_0), \tag{21}$$

producing the nonlinear difference equation

$$g(t) = kg(r),\ t = r + 2 + (1+k)\frac{(\gamma+1)}{2}g(r). \tag{22}$$

Consider the initial value problem:

$$g(t_0) = M\psi(t_0), \qquad 0 \le t_0 < 2, \tag{23}$$

where $M \ll 1$ is the Mach number. Then (22) implies

$$g(t_{n+1}) = kg(t_n) = k^{n+1}M\psi(t_0), \tag{24}$$

where

$$t_{n+1} = t_0 + 2n + \frac{k+1}{1-k}[1-k^{n+1}]\frac{(\gamma+1)}{2}]M\psi(t_0).$$

A shock forms when $\frac{dg(t_{n+1})}{dt_{n+1}} \to \infty$. But (24) implies that

$$\frac{dg(t_{n+1})}{dt_{n+1}} = k^{n+1}M\psi'(t_0)\frac{dt_0}{dt_{n+1}},$$

so that a shock will form when $\frac{dt_{n+1}}{dt_0} \to 0$ or

$$0 = \frac{dt_{n+1}}{dt_0} = 1 + \frac{k+1}{1-k}[1-k^{n+1}]\frac{(\gamma+1)}{2}M\psi'(t_0). \tag{25}$$

For a shock to form when $|k| < 1$ we require

$$\left|\frac{(\gamma+1)}{2}M\psi'(t_0)\right| > \left|\frac{1-k}{1+k}\right| = i \tag{26}$$

for some t_0 with $0 < t_0 < 2$. This says that there is a critical initial acceleration level, $|M\psi'(t_{cr})| = 2i/(\gamma+1)$, and a shock will only form for an initial acceleration above this level. Otherwise the signal is damped out before a shock can form. When $k = 1$ $(i = 0)$ there is always a shock; see Mortell & Varley [7].

5. A self-sustained oscillation

For the case $|k| > 1$ the signal is amplified geometrically. This is an example of a system that is linearly unstable to perturbations about the initial state. However, within nonlinear theory a shock always forms. Since shocks dissipate energy, this raises the possibility of a balance between the energy flowing in across the boundary and the shock dissipation.

As in the previous section, the linear solution satisfies the difference equation $g(t) = kg(t-2)$ so that in linear theory

$$g(t+2n) = k^n g(t), \; k > 1. \tag{27}$$

Hence g grows like k^n and is linearly unstable.

Within nonlinear theory the solution is governed by (22). Shocks form and the signal will eventually evolve to an $N-$ wave, or a series of $N-$waves, so will have a linear slope passing through the zeros of $g(t)$. The solution is found through the use of critical points of the nonlinear difference equation, see Seymour & Mortell [8] for details. We define a critical point, $t = r = t_c$, as a location where $g(t) = g(r)$; here $g(t_c) = 0$. If (22) is differentiated with respect to r and evaluated at $t = t_c$ we obtain a quadratic equation for $\lambda = g'(t_c)$:

$$\lambda(\lambda - \frac{2(k-1)}{(k+1)(\gamma+1)}) = 0. \tag{28}$$

The root $\lambda = 0$ yields the trivial solution $g(t) = 0$, while the nonzero root gives the periodic solution

$$g(t) = \frac{2(k-1)}{(k+1)(\gamma+1)}(t - t_c), \; g(t+2) = g(t), \tag{29}$$

with the mean condition $\int_0^1 g(y)dy = 0$. Substitution of (29) into (22) shows that this is an exact solution of the difference equation. The final periodic state is therefore piecewise linear, passing through alternate zeros of the initial signal function and joined by shocks which are of constant strength, $\frac{4(k-1)}{(k+1)(\gamma+1)}$, see Chu [9] and Mortell & Seymour [10].

6. Damped resonance in a closed tube

The general solution is

$$u(\alpha, \beta) = f(\alpha) + g(\beta), \; e = f(\alpha) - g(\beta), \tag{30}$$

where in linear theory $\alpha = t - x$, $\beta = t + x - 1$. The boundary condition at $x = 0$, corresponding to an outflow of energy ($i > 0$), is

$$e(0,t) = -iu(0,t), \tag{31}$$

so that

$$f(t) = -kg(t-1), \; k = \frac{i-1}{i+1}, \quad 0 \le i < \infty, \; k \le 1. \tag{32}$$

At $x = 1$ there is periodic forcing

$$u(1,t) = Mh(\omega t), \tag{33}$$

where h is periodic with unit period, $\int_0^1 h(y)dy = 0$ and M is the Mach number. From the equations of motion, for a periodic response we require $\int_0^1 g(y)dy = 0$.

Linear difference equation

Let $y = \omega t$, then $g(y)$ satisfies

$$g(y) - kg(s) = Mh(y), \; y = s + 2\omega. \tag{34}$$

At $\omega = \frac{1}{2}$ this becomes

$$g(y) - kg(y-1) = Mh(y), \tag{35}$$

so there is no solution with unit period when $k = 1$. To consider problems at frequencies around resonance, define the detuning from resonance $\frac{1}{2}\Delta$ by

$$\omega = \frac{1}{2}(1+\Delta), \tag{36}$$

then

$$g(y) - kg(s) = Mh(y), \; y = s + 1 + \Delta, \tag{37}$$

and the **nonlinear travel time** is

$$y = s + 1 + \Delta + \omega(1+k)\frac{(\gamma+1)}{2}g(s). \tag{38}$$

We define $G(y)$ and $H(y)$ as

$$G(y) = \Delta + \omega(1+k)\frac{(\gamma+1)}{2}g(y), \; MH = \mu\Delta + \omega(1+k)\frac{(\gamma+1)}{2}Mh, \tag{39}$$

where $0 < \mu = 1 - k < 1$ and now $\int_0^1 G(y)dy = \Delta$.

Then the **nonlinear difference equation** is

$$G(y) - kG(s) = MH(y), \; y = s + G(s). \tag{40}$$

on noting $g(y+1) = g(y)$. This is the *Dissipative Standard Mapping*.

Assuming $M \ll 1, |G| \ll 1$ and $|G'| \ll 1$, (40) reduces to the o.d.e.

$$G(y)G'(y) + \mu G(y) = MH(y), \tag{41}$$

subject to

$$\int_0^1 G(y)dy = \Delta, \; G(y+1) = G(y), \tag{42}$$

since $\int_0^1 g(y)dy = 0$, to determine the solution, see Seymour & Mortell [11] and Chester [12].

7. Evolution of damped resonance

The canonical equation is (40), where M is the Mach number of the input. For $M \ll 1$ we write

$$k = 1 - k_1 M^{1/2} \text{ and } \omega = \frac{1}{2}(1 + M^{1/2}\Delta_1), \tag{43}$$

so that (40) becomes

$$G(y) - (1 - k_1 M^{1/2})G(s) = MH(y), \; y = s + 1 + G(s). \tag{44}$$

We assume a long-time multiple scale expansion of the form

$$G(s) = M^{1/2}G_0(s,\tau) + MG_1(s,\tau) + \ldots.., \; \tau = M^{1/2}s, \tag{45}$$

then at $O(M^{1/2})$,

$$G_0(s+1,\tau) - G_0(s,\tau) = 0. \tag{46}$$

At $O(M)$, to eliminate growth terms, G_0 satisfies the nonlinear kinematic wave equation with no secular growth:

$$\frac{\partial G_0(s,\tau)}{\partial \tau} + G_0(s,\tau)\frac{\partial G_0(s,\tau)}{\partial s} + k_1 G_0(s,\tau) = H(s), \tag{47}$$

with $G_0(s, 0) = \Delta_1$.

If $k_1 = 0$, the linear equation gives

$$G(s) = M^{1/2} G_0(s,\tau) = \tau M^{1/2} H(s) = MsH(s), \tag{48}$$

which predicts initial linear growth. So the p.d.e. (47) is uniformly valid, see Cox & Mortell [13].

8. Resonance between concentric spheres

We consider sound waves in a spherical shell, $R_a < R < R_b$, generated by a periodically pulsating boundary at R_b with period τ_p. We nondimensionalize length with $L = R_b - R_a$, time with L/c_0, where $c_0 = \sqrt{\gamma p_0/\rho_0}$ is the ambient sound speed, and particle velocity with c_0. Then define $r_a = R_a/L$ and $r_b = R_b/L$ so $r_b - r_a = 1$.

The dimensionless governing equations are:

$$u_t + uu_r + b^{-1} cc_r = 0, \; c_t + uc_r + bc(u_r + 2u/r) = 0, \; r_a < r < r_b, \tag{49}$$

where $\omega = L/(\tau_p c_0)$, $u(r,t)$ is the particle velocity, $c(r,t)$ is the (dimensionless) sound speed and $b = (\gamma - 1)/2$ is a constant.

The rigid shell boundary condition on $r = r_a$ is

$$u(r_a, t) = 0, \tag{50}$$

while the oscillating boundary condition on $r = r_b$ is

$$u(r_b, t) = M \sin(2\pi\omega t), \tag{51}$$

where $M = u_0/c_0$ is the Mach number of the input.

The spherical wave general solution in linear theory is

$$u = \frac{1}{r}[f(\alpha) + g(\beta)], \; \alpha = t - (r - r_a), \; \beta = t + (r - r_b), \tag{52}$$

Following a wave leaving $r = r_b$ at time $t = t_0$, reflected from r_a at $t = t_1$ and then again at $r = r_b$ at $t = t_2$, the linear characteristics give the linear travel time

$$t_2 - t_0 = 2. \tag{53}$$

The boundary conditions then imply that

$$g(t_2) - g(t_0) = r_b M \sin(2\pi\omega t_2). \tag{54}$$

This is exactly as a straight tube with M replaced by Mr_b.

The nonlinear travel time, found by a nonlinear geometric acoustics expansion for the characteristics, see Whitham [3], with the restriction $L/R_a \ll 1$, or $1/r_a \ll 1$, is

$$t_2 - t_0 = 2 + g(t_0)\ln(r_b/r_a)(1+\gamma). \tag{55}$$

The geometry enters through the term $\ln(r_b/r_a)$, and the result (55) should be compared with (16) for a straight tube.

Then (54) and (55) determine the nonlinear resonant oscillations when ω lies in the neighbourhood of resonance, $\omega = \frac{1}{2}$. At resonance

$$g(t)g'(t) = \frac{r_b}{1+\gamma}\ln(r_b/r_a)M\sin(\pi t), \quad (56)$$

$$g(t+2) = g(t), \text{ and } \int_0^2 g(y)dy = 0, \quad (57)$$

determine the solution, see Seymour, Mortell & Amundsen [14] and Galiev & Panova [15].

9. Resonance between concentric cylinders

We consider sound waves in a cylindrical shell, $R_a < R < R_b$, generated by a periodically pulsating boundary at R_b with period τ_p. In this case the linear geometric acoustics approximation, $L/R_a \ll 1$, produces the linear solution

$$u = \frac{1}{\sqrt{r}}[f(\alpha) + g(\beta)]. \quad (58)$$

This is used to calculate the linear difference equation and the nonlinear characteristics from which the nonlinear travel time is calculated. Hence the corrected nonlinear characteristics are

$$\alpha = t - (r - r_a) + f(\alpha)[\sqrt{r} - \sqrt{r_a}](1+b) + ..., \ \alpha = t \text{ on } r = r_a \quad (59)$$

$$\beta = t + (r - r_b) + g(\beta)[\sqrt{r} - \sqrt{r_b}](1+b) + ..., \ \beta = t \text{ on } r = r_b, \quad (60)$$

where $b = (\gamma - 1)/2$, see Whitham [3].

The boundary conditions are

$$u(r_a, t) = 0, \ u(r_b, t) = M\sin(2\pi\omega t), \ M \ll 1.$$

Then the linear difference equation is

$$g(t_2) - g(t_0) = \sqrt{r_b}M\sin(2\pi\omega t_2), \quad (61)$$

where $t_2 - t_0 = 2$ is the linear travel time. This is exactly as the previous linear (straight tube) equation with M replaced by $M\sqrt{r_b}$. There is no periodic solution for $\omega = 1/2$, the resonant frequency.

The approximate nonlinear characteristics give the nonlinear travel time as

$$t_2 - t_0 = 2 + 2g(t_0)[\sqrt{r_b} - \sqrt{r_a}](1+\gamma), \quad (62)$$

which can be compared with (16) for a straight tube.

The nonlinear solution is found from (61) and (62) as in (56) and (57), see Seymour, Mortell & Amundsen [14].

Acknowledgments

The research reported in this paper was supported in part by NSERC Discovery Grant A9117 (BRS).

10. References

[1] Saenger, R A and Hudson, G E 1960 Periodic shock waves in resonating gas columns *Journal of the Acoustical Society of America* **32 (8)** 961.

[2] Whitham, G B 1952 The flow pattern of a supersonic projectile *Communications on pure and applied mathematics* **5 (3)** 301.

[3] Whitham, G B 1974 *Linear and nonlinear waves* John Wiley & Sons New York.

[4] Landau, L D 1945 On shock waves at large distances from the place of their origin *Soviet Journal of Physics* **9** 496.

[5] Lighthill, M J 1961 A technique for rendering approximate solutions to physical problems uniformly valid. *Zeitschrift für Flugwissenschaften* **9** 267.

[6] Lax, P D 1964 Development of singularities in solutions of nonlinear hyperbolic partial differential equations *Journal of Mathematical Physics* **5** 611.

[7] Mortell, M P and Varley E 1970 Finite amplitude waves in bounded media. nonlinear free vibrations of an elastic panel *Proceedings of the Royal Society of London. A. Mathematical and Physical Sciences* **318** 169.

[8] Seymour, B R and Mortell, M P 1980 A finite-rate theory of resonance in a closed tube: discontinuous solutions of a functional equation. *Journal of Fluid Mechanics* **99** 365.

[9] Chu, B-T 1963 Analysis of a self-sustained thermally driven nonlinear vibration *Physics of Fluids* **6 (11)** 1638.

[10] Mortell, M P and Seymour B R 1973 The evolution of a self-sustained oscillation in a nonlinear continuous system *Journal of Applied Mechanics* **40 (1)** 53.

[11] Seymour, B R and Mortell M P 1973 Resonant acoustic oscillations with damping: small rate theory *Journal of Fluid Mechanics* **58** 353.

[12] Chester, W 1964 Resonant oscillations in closed tubes *Journal of Fluid Mechanics* **18** 44.

[13] Cox, E A and Mortell M P 1989 The evolution of resonant acoustic oscillations with damping *Proceedings of the Royal Irish Academy* **89A(2)** 147.

[14] Seymour, B R, Mortell M P and Amundsen D E 2012 Asymptotic solutions for shocked resonant acoustic oscillations between concentric spheres and coaxial cylinders *Physics of Fluids* **24 (2)**.

[15] Galiev, S U and Panova O P 1995 Periodic shock waves in a spherical resonator *Strength of Materials* **27 (10)** 602.

An asymptotic analysis of the laminar-turbulent transition of yield stress fluids in pipes

Tim G. Myers
Centre de Recerca Matemàtica, Barcelona, Spain
E-mail: tmyers@crm.cat

Sarah L. Mitchell
University of Limerick, Ireland
E-mail: sarah.mitchell@ul.ie

Paul Slatter
ATC Williams, Mordialloc, VIC, 3195, Australia
E-mail: pauls@atcwilliams.com.au

Abstract. The work in this paper concerns the axisymmetric pipe flow of a Herschel-Bulkley fluid, with the aim of determining a relation between the critical velocity (defining the transition between laminar and turbulent flow) and the pipe diameter in terms of the Reynolds number Re_3. The asymptotic behaviour for large and small pipes is examined and simple expressions for the leading order terms are presented. Results are then compared with experimental data. A nonlinear regression analysis shows that for the tested fluids the transition occurs at similar values to the Newtonian case, namely in the range $2100 < Re_3 < 2500$.

1. Introduction

Yield stress fluids are transported through pipes in a number of different industries and scenarios, such as the transport of crude oil, mining slurries, liquid food, concrete, bio-fluids and drilling fluids or sewage sludge, see [6, 10, 12] for example. In certain cases, many mining slurries for example, the yield stress behaviour is due to a high concentration of particles suspended in the carrier fluid. To prevent blockage of the pipe a sufficiently high velocity must be maintained, so that the flow remains turbulent, which prevents the particles from settling. However, the larger the velocity, the more expensive the operation and consequently designers and operators attempt to restrict the flow to a level only slightly above the laminar-turbulent transition point. Obviously this is a risky strategy, since errors can lead to expensive pipe blockage, and so it is critical that this transition point is accurately identified. In the following work we will focus on the flow of suspensions using the Herschel-Bulkley model. Although the Herschel-Bulkley model is not appropriate for the laminar flow of suspensions (even in the turbulent transition region most particles will settle) we will still employ it in order to derive the appropriate mathematical model. The expression for the Reynolds number in terms of the flow parameters will then be compared to experimental data to identify the laminar-turbulent transition.

The laminar-turbulent transition point for Newtonian fluids is well known and defined in terms of the Reynolds number

$$Re = \frac{\rho V D}{\eta} , \tag{1}$$

where ρ is the density, V the average velocity, D the pipe diameter and η the dynamic viscosity. Transition typically occurs for $Re \sim 2300$, although values between 1760 and 2300 are quoted in Kerswell [7]. This issue is discussed in great detail in the survey of Eckhardt *et al.* [3]. For yield stress fluids the matter is even less clear; one problem being that there is not a unique definition of the Reynolds number. Metzner & Reed [8] proposed a generalised Reynolds number

$$Re_{MR} = \frac{8\rho V^2}{\tau_{0_l}} , \tag{2}$$

where τ_{0_l} is the shear stress at the wall under laminar flow conditions. Wasp *et al.* [21] define a Reynolds number for a Bingham fluid (i.e. $n = 1$) based on the Hedström number

$$Re_W = 1500 \left(1 + \sqrt{1 + \frac{\rho D^2 \tau_y}{4500 K^2}} \right) , \tag{3}$$

where τ_y is the yield stress and K the consistency index. A common generalisation of the Newtonian Reynolds number, Re_w, involves defining the viscosity at the wall, so $\eta = \eta_w$ in (1), where η_w is the ratio of the shear stress and shear rate at the wall, see [2, 11, 19] for example. For a yield stress fluid flowing in a pipe, Slatter [1, 15] proposes a formulation that takes this concept even further. Instead of focussing on the flow solely at the pipe wall they include the flow in the annular region surrounding the central plug. The plug flow is neglected based on the premise that this does not form part of the sheared region, and is not behaving as a fluid. The analysis leads to what is termed Re_3

$$Re_3 = \frac{8\rho V_a^2}{\tau_y + K\left(\frac{8V_a}{D_a}\right)^n} , \tag{4}$$

where V_a is the average fluid velocity in the sheared annulus and D_a is the difference between the pipe diameter and the plug diameter.

Not only is the appropriate form of the Reynolds number subject to debate, there is also confusion over the value at which the transition occurs; although this is most likely fluid dependent. In [4] the flow of a Laponite fluid is investigated, using a Herschel-Bulkley model, and their experiments indicate a transition to turbulence for $Re_w \sim 3400$. Rudman *et al.* [11] suggest that as the power law exponent decreases, so the flow moves further away from Newtonian, and the transitional value of Re_w increases. Their experiments indicate that transition occurs for $Re_w \in [1300, 3000]$.

Draad *et al.* investigate polymer solutions which are modelled well by the Carreau viscosity relation (i.e. there is no yield stress). They conclude that for flow in relatively large pipes (compared to the polymer length-scale) the transition will occur around the same value as for a Newtonian solution, $Re_w \sim 2300$. This result may be related to Re_3 via the following argument. In the limit $\tau_y \to 0$ the annular velocity and diameter reduce to the average velocity and pipe diameter, $V_a \to V$, $D_a \to D$ and

$$Re_3 \to \frac{8^{1-n}\rho V^{2-n} D^n}{K} , \tag{5}$$

where n is the exponent in the fluid viscosity definition. The viscosity at the wall η_w is

$$\eta_w = \frac{K^{1/n}\tau_0}{(\tau_0 - \tau_y)^{1/n}} .$$

(Note, this may be calculated using the expression for velocity given in the following section). Combining this with the expression for V (also given in the following section) and setting $\tau_y = 0$, we determine

$$Re_w = \left(\frac{6n+2}{n}\right)^{1-n} \frac{\rho V^{2-n} D^n}{K} . \quad (6)$$

The expressions for Re_w and Re_3 only differ by a factor $f(n) = (8n/(6n+2))^{1-n}$. For $n \in [0.5, 1]$ the factor $f \in [0.9, 1]$ and consequently we expect that, provided $\tau_y \sim 0$ and $n \in [0.5, 1]$, the transition value for Re_3 should also be close to 2300. The experimental results of Slatter [16] confirm this by showing that a choice of Re_3 anywhere in the range $[2100, 2400]$ provides accurate results for the transition. Observe that the experimental data quoted in Table 1 in the appendix have $n > 0.59$, whereas the results of Draad *et al.* [2] have $n > 0.61$.

The principal objective of Slatter's work [12, 13, 14, 15, 16] was to establish a simple, single criterion for transition such as exists for Newtonian pipe flow, e.g. $Re = 2300$ [16]. This obviously requires an appropriate definition of the Reynolds number. In §4 we show that $Re_3 \in [2100, 2500]$ provides good agreement with experimental data and consequently is an appropriate non-dimensional parameter to categorize the behaviour of a Herschel-Bulkley fluid.

The Reynolds number depends on both the velocity and pipe diameter. To maintain turbulent flow, for a given pipe diameter, the mean fluid velocity must be kept above a critical value V_c. One focus of this work is to find a simple relation between the critical mean velocity, V_c, and the pipe diameter, D, for large pipes. For sufficiently large pipes, $D = \mathcal{O}(1)$ m, experimental observations indicate that the critical velocity becomes independent of the pipe diameter [12]. In this limit, dimensional analysis shows that the dependence of V_c then takes the form

$$V_c = C(n)\sqrt{\frac{\tau_y}{\rho}} . \quad (7)$$

However, the dimensional analysis cannot determine the form of the coefficient $C(n)$, which must at present be approximated numerically for a given fluid. One of our goals is therefore to find an analytical expression for $C(n)$. Our other goal is to show that Re_3 is an appropriate non-dimensional number to characterize the flow of a yield-stress fluid. Once this has been established we can then determine the correct value of Re_3 that predicts the transition to turbulence.

The structure of the paper is as follows. In §2 we quote standard results to describe the flow of a Herschel-Bulkley fluid in a pipe and consequently determine V_a and D_a which are required in the expression for Re_3 in (4). These results are then used in §3 to determine different asymptotes for the ratio of yield stress to wall shear stress: these can then be translated onto the critical velocity-pipe diameter diagram. In particular we find an analytical expression for $C(n)$. In §4 we compare our analytical model with the experimental data of Slatter [16] and use a nonlinear regression analysis to determine the best choice of Re_3.

2. Governing equations for the flow

Under conditions of fully developed, steady-state, one dimensional laminar pipe flow, a yield-stress fluid will shear in the annular region next to the pipe wall. Near the centre of the pipe, where the applied shear stresses are less than the yield stress, there will be a solid, unsheared plug which is carried along by the applied pressure gradient at the maximum fluid velocity: this

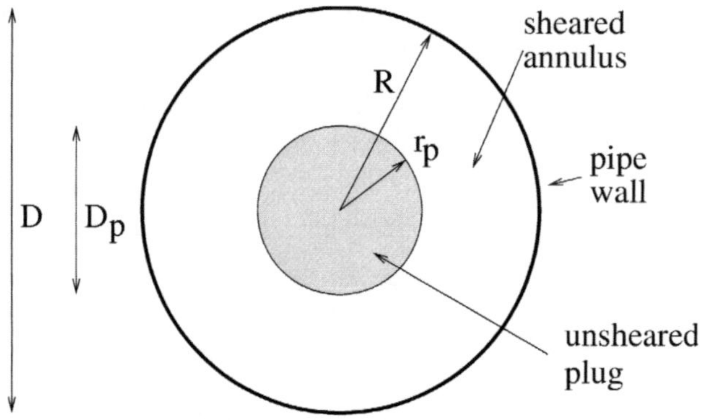

Figure 1. *Steady, unidirectional flow geometry.*

is depicted in Figure 1. The velocity is given by $\mathbf{u} = (0, w(r))$, where w is the velocity along the axis.

Velocity and shear stress relations for axisymmetric pipe flow are well known. For unidirectional flow, the shear stress in the annular region for a Herschel-Bulkley fluid is defined by

$$\tau = -\tau_y - K\left(-\frac{\partial w}{\partial r}\right)^n , \qquad |\tau| > \tau_y , \tag{8}$$

where K is the consistency index and τ_y the yield stress. In the plug region $0 \leq r \leq r_p$, where $|\tau| < \tau_y$, the velocity is constant $w_r = 0$. In the annular region $r_p \leq r \leq R$ the velocity profile is

$$w = \frac{2n}{(n+1)p_z K^{1/n}}\left[\left(-\frac{p_z r}{2} - \tau_y\right)^{(n+1)/n} - \left(-\frac{p_z R}{2} - \tau_y\right)^{(n+1)/n}\right] , \tag{9}$$

where p_z is the pressure gradient. Since $\tau = -\tau_y$ at $r = r_p$ the plug radius can be expressed as

$$r_p = -\frac{2\tau_y}{p_z} . \tag{10}$$

Substituting for $r = r_p$ in (9) determines the plug velocity

$$w_p = -\frac{2n}{(n+1)p_z K^{1/n}}\left(-\frac{p_z R}{2} - \tau_y\right)^{(n+1)/n} . \tag{11}$$

Equation (4) for Re_3 requires expressions for the annular velocity V_a and the diameter of the annular region D_a. The latter quantity is simply

$$D_a = 2(R - r_p) , \tag{12}$$

where r_p is defined by equation (10). The annular velocity $V_a = Q_a/A_a$ requires an expression for the annular flux which is the difference between the total flux and the flux of the plug region

$$Q_a = Q - Q_p = Q - w_p \pi r_p^2 , \tag{13}$$

and so

$$V_a = \frac{Q - w_p \pi r_p^2}{\pi(R^2 - r_p^2)} \,. \tag{14}$$

The total flux is $Q = VA = V\pi R^2$ where V is the mean velocity in the pipe:

$$V = \frac{2\pi}{\pi R^2}\left[\int_0^{r_p} r w_p \,\mathrm{d}r + \int_{r_p}^{R} r w \,\mathrm{d}r\right] \tag{15}$$

$$= \frac{Dn}{2K^{1/n}\tau_0^3}(\tau_0 - \tau_y)^{(n+1)/n}\left[\frac{(\tau_0-\tau_y)^2}{3n+1} + \frac{2\tau_y(\tau_0-\tau_y)}{2n+1} + \frac{\tau_y^2}{n+1}\right] . \tag{16}$$

From the definition of the stress we find the ratio $r_p/R = \tau_y/\tau_0$. Using this and the definitions of V and w_p we may write V_a in (14) as

$$V_a = \frac{V\pi R^2 - w_p \pi r_p^2}{\pi(R^2 - r_p^2)} = \frac{V\tau_0^2 - w_p\tau_y^2}{\tau_0^2 - \tau_y^2} = \frac{Dn}{2K^{1/n}}\frac{(\tau_0-\tau_y)^{(n+1)/n}}{\tau_0(\tau_0+\tau_y)}\left[\frac{\tau_0-\tau_y}{3n+1} + \frac{2\tau_y}{2n+1}\right] . \tag{17}$$

Similarly D_a from (12) becomes

$$D_a = \frac{D(\tau_0 - \tau_y)}{\tau_0} \,. \tag{18}$$

Substitution of the above expressions for V_a and D_a, equations (17, 18), into the definition of Re_3 (4) therefore gives

$$Re_3 = \frac{\rho D^2}{8K^{2/n}}\frac{(\tau_0-\tau_y)^{2(n+1)/n}\tau_0^{-2}(\tau_0+\tau_y)^{-2}\left[a(\tau_0-\tau_y)+2b\tau_y\right]^2}{\tau_y + (\tau_0-\tau_y)(\tau_0+\tau_y)^{-n}\left[a(\tau_0-\tau_y)+2b\tau_y\right]^n} \,, \tag{19}$$

while the expression for V in (16) becomes

$$V = \frac{D}{8K^{1/n}}(\tau_0-\tau_y)^{(n+1)/n}\tau_0^{-3}\left[a(\tau_0-\tau_y)^2 + 2b\tau_y(\tau_0-\tau_y) + c\tau_y^2\right] , \tag{20}$$

where, to simplify notation, we set

$$a = \frac{4n}{3n+1} \,, \qquad b = \frac{4n}{2n+1} \,, \qquad c = \frac{4n}{n+1} \,. \tag{21}$$

3. Asymptotic analysis

Now that Re_3 and the average velocity (19, 20) have been calculated we can determine the relation between the Reynolds number, the mean velocity V and the pipe diameter D. The aim of this work is to find the critical velocity, $V = V_c$, above which the flow is turbulent. This should occur at a specific value of Re_3. The following analysis leads to a relation of the form $V = f(Re_3, D)$ and by choosing a specific value of Re_3 we determine the relation between V_c and D. We can then show, from a nonlinear regression analysis using experimental data [12, 16], that the transition to turbulent flow occurs in the range $2100 < Re_3 < 2500$.

We find it convenient to introduce the parameter $\mathcal{A} = \tau_0/\tau_y = R/r_p$, which is the ratio of wall shear stress to yield stress or equivalently the pipe radius to the plug radius. In the following we will assume that the fluid properties are fixed and so control $\mathcal{A}$ by varying the pipe diameter. In the small D limit the high shear rates required to move the fluid will confine the plug to a small region near $r = 0$. This means that $A = R/r_p \gg 1$ as $D \to 0$. Experimental results, see [12] for example, show that for large D the critical velocity approaches a constant value. This is confirmed by dimensional analysis in [12] for the case $n < 1$. For the large D limit, letting $D \to \infty$ in equation (20) shows that to prevent $V \to \infty$ then $\tau_0 \to \tau_y$ (in fact $\tau_o - \tau_y \propto D^{-n/(n+1)}$) and hence $\mathcal{A} \to 1^+$.

In the following analysis it is convenient to set $\tau_0 = \mathcal{A}\tau_y$, for $\mathcal{A} > 1$, and then the large D limit corresponds to $\mathcal{A} \approx 1$, whereas the small D limit corresponds to $\mathcal{A} \gg 1$. Substitution of $\tau_0 = \mathcal{A}\tau_y$ into (19) and (20) gives

$$Re_3 = \frac{\rho D^2 \tau_y^{2/n-1}}{8K^{2/n}} \frac{(\mathcal{A}-1)^{2(n+1)/n}\mathcal{A}^{-2}(\mathcal{A}+1)^{-2}[a(\mathcal{A}-1)+2b]^2}{1+(\mathcal{A}-1)(\mathcal{A}+1)^{-n}[a(\mathcal{A}-1)+2b]^n}, \tag{22}$$

$$V = \frac{D\tau_y^{1/n}}{8K^{1/n}}(\mathcal{A}-1)^{(n+1)/n}\mathcal{A}^{-3}\Big[a(\mathcal{A}-1)^2 + 2b(\mathcal{A}-1) + c\Big]. \tag{23}$$

We now consider the cases $\mathcal{A} \approx 1$ and $\mathcal{A} \gg 1$ separately.

3.1. *The $\mathcal{A} \approx 1$ (large D) asymptote*

Since $\mathcal{A} \approx 1$ (but strictly $\mathcal{A} > 1$ to allow some flow) we write $\mathcal{A}$ as the following expansion:

$$\mathcal{A} - 1 = \alpha_0\epsilon + \alpha_1\epsilon^2 + \alpha_2\epsilon^3 + \ldots, \tag{24}$$

where $\alpha_i = \mathcal{O}(1)$, $\alpha_0 > 0$ and $0 < \epsilon \ll 1$. Substitution of (24) into (22) leads to

$$\begin{aligned} Re_3 = \frac{\rho D^2 \tau_y^{2/n-1}}{8K^{2/n}} b^2 (\alpha_0\epsilon)^{2(n+1)/n} \Bigg[1 &+ \left\{ \frac{a}{b} - 3 - b^n + \frac{2(n+1)}{n}\frac{\alpha_1}{\alpha_0^2} \right\} \alpha_0\epsilon \\ &+ \left\{ \frac{a^2}{4b^2} + \frac{23}{4} - \frac{3a}{b} + \frac{nb^n}{2}\left(1 - \frac{a}{b}\right) - \left(\frac{3n+2}{n}\left(3 - \frac{a}{b}\right) + b^n\right)\frac{\alpha_1}{\alpha_0^2} \right. \\ &\left. + \frac{(n+1)(n+2)}{n^2}\frac{\alpha_1^2}{\alpha_0^4} + \frac{2(n+1)}{n}\frac{\alpha_2}{\alpha_0^3} \right\} (\alpha_0\epsilon)^2 + \ldots \Bigg], \end{aligned} \tag{25}$$

and in (23) we now have

$$\begin{aligned} V = \frac{D\tau_y^{1/n}}{8K^{1/n}} c(\alpha_0\epsilon)^{(n+1)/n} \Bigg[1 &+ \left(\frac{2b}{c} - 3 + \frac{n+1}{n}\frac{\alpha_1}{\alpha_0^2} \right) \alpha_0\epsilon \\ &+ \left\{ 6 + \frac{a}{c} - \frac{6b}{c} + \frac{2n+1}{n}\left(\frac{2b}{c} - 3\right)\frac{\alpha_1}{\alpha_0^2} + \frac{n+1}{2n^2}\frac{\alpha_1^2}{\alpha_0^4} + \frac{n+1}{n}\frac{\alpha_2}{\alpha_0^3} \right\} (\alpha_0\epsilon)^2 + \ldots \Bigg]. \end{aligned} \tag{26}$$

The expression for $\alpha_0\epsilon$ is found from the leading order term in (25) and so

$$\alpha_0\epsilon = \left(\frac{8Re_3 K^{2/n}}{\rho\tau_y^{(2-n)/n}} \right)^{n/2(n+1)} (bD)^{-n/(n+1)}. \tag{27}$$

Setting the higher order correction terms to zero determines α_1, α_2. Substitution of $\alpha_0, \alpha_1, \alpha_2$ into (26) gives V as a function of D.

If we simply consider the leading order term and substitute for $\alpha_0\epsilon$, via (27), into (26) then V becomes

$$V \sim \frac{c}{b}\sqrt{\frac{Re_3\tau_y}{8\rho}} = C(n)\sqrt{\frac{\tau_y}{\rho}}, \qquad C(n) = \frac{2n+1}{n+1}\sqrt{\frac{Re_3}{8}}. \tag{28}$$

As discussed in the introduction, finding the expression in (28) was one of the main goals of the original analysis [9]. This equation identifies the asymptote as $D \to \infty$ and the form of the fluid constant $C(n)$. However, including the correction terms $\mathcal{O}[\epsilon]$ and $\mathcal{O}[\epsilon^2]$ in (26) means that we can find the form of the curve for smaller D and so get closer to the transition between the small and large D asymptotes. This can be seen in more detail in the Results section below. Note that the condition $\alpha_0\epsilon \ll 1$ provides a limit for the pipe diameter

$$D \gg \frac{1}{b}\left(\frac{8Re_3 K^{2/n}}{\rho\tau_y^{(2-n)/n}}\right)^{1/2} . \tag{29}$$

The large D expansion can also be derived by considering the expressions for Re_3 and V in (19) and (20) directly. Since $\tau_0 \to \tau_y$ in this limit, a high proportion of the slurry will be at the yield stress and consequently the plug occupies most of the pipe. Experiments also show that the transition velocity V_c becomes independent of D as $D \to \infty$ [12]. From examining (20) it is clear that this can only occur if $(\tau_0 - \tau_y)^{(n+1)/n} = \mathcal{O}[D^{-1}]$. We could therefore look for an expansion of $\tau_0 - \tau_y$ in terms of the small parameter $\epsilon = D^{-n/(n+1)}$, namely

$$\tau_0 - \tau_y = \alpha_0 D^{-n/(n+1)} + \alpha_1 D^{-2n/(n+1)} + \mathcal{O}((D^{-n/(n+1)})^3) = \alpha_0\epsilon + \alpha_1\epsilon^2 + \mathcal{O}(\epsilon^3) .$$

This method yields the same expansions for Re_3 and V and is discussed in more detail in [9].

3.2. The large $\mathcal{A}$ (small D) asymptote

In this case $\mathcal{A} \gg 1$ and so we consider an expansion in terms of the small parameter $\mathcal{B} = 1/\mathcal{A}$. Rewriting Re_3 and V from (22) and (23) in terms of $\mathcal{B}$ gives

$$Re_3 = \frac{\rho D^2 \tau_y^{2/n-1}}{8K^{2/n}} \frac{(1-\mathcal{B})^{2(n+1)/n}\mathcal{B}^{1-2/n}(1+\mathcal{B})^{-2}[a(1-\mathcal{B}) + 2b\mathcal{B}]^2}{\mathcal{B} + (1-\mathcal{B})(1+\mathcal{B})^{-n}[a(1-\mathcal{B}) + 2b\mathcal{B}]^n} , \tag{30}$$

$$V = \frac{D\tau_y^{1/n}}{8K^{1/n}}\mathcal{B}^{-1/n}(1-\mathcal{B})^{(n+1)/n}\Big[a(1-\mathcal{B})^2 + 2b\mathcal{B}(1-\mathcal{B}) + c\mathcal{B}^2\Big] . \tag{31}$$

Suppose $\mathcal{B}$ is expanded as

$$\mathcal{B} = \beta_0\epsilon + \beta_1\epsilon^2 + \beta_2\epsilon^3 + \ldots , \tag{32}$$

where $\beta_i = \mathcal{O}(1)$, $\beta_0 > 0$ and $0 < \epsilon \ll 1$. Substitution into (30) and (31) leads to expansions

$$\begin{aligned} Re_3 = {} & \frac{\rho D^2\tau_y^{2/n-1}}{8K^{2/n}} a^{2-n}(\beta_0\epsilon)^{1-2/n}\Bigg[1 + \left\{\frac{2b(2-n)}{a} - \frac{5n+2-2n^2}{n} - a^{-n} + \frac{n-2}{n}\frac{\beta_1}{\beta_0^2}\right\}\beta_0\epsilon \\ & \left\{\frac{n-n^2+2}{2}\left(-1+\frac{2b}{a}\right)^2 + \frac{n^3+n^2-8n-4}{n}\left(-1+\frac{2b}{a}\right)\right. \\ & + \left(\frac{2b(4n-n^2-4}{an} + \frac{2n^3-11n^2+8n+4}{n^2} - a^{-n}\right)\frac{\beta_1}{\beta_0^2} \\ & \left. - \frac{3n^3+n^4-16n^2-14n-4}{2n^2} - \frac{n-2}{n^2}\frac{\beta_1^2}{\beta_0^4} + \frac{n-2}{n}\frac{\beta_1}{\beta_0^2}\right\}(\beta_0\epsilon)^2 + \ldots\Bigg] , \end{aligned} \tag{33}$$

and

$$V = \frac{D\tau_y^{1/n}}{8K^{1/n}} a(\beta_0\epsilon)^{-1/n}\Bigg[1 + \left(\frac{2b}{a} - 2 - \frac{n+1}{n} - \frac{1}{n}\frac{\beta_1}{\beta_0^2}\right)\beta_0\epsilon$$

$$+\left\{1+\frac{c}{a}-\frac{2b}{a}+\frac{n-1}{2n^2}-\frac{2(n+1)}{n}\left(-1+\frac{b}{a}\right)\right.$$
$$\left.+\left(\frac{2b}{a}-2+\frac{1-n^2}{n}-\frac{2}{n}\left(-1+\frac{b}{a}\right)\right)\frac{\beta_1}{\beta_0^2}+\frac{n+1}{2n^2}\frac{\beta_1^2}{\beta_0^4}-\frac{1}{n}\frac{\beta_2}{\beta_0^3}\right\}(\beta_0\epsilon)^2+...\Bigg]. \quad (34)$$

From the leading order term in (33) we obtain an expression for $\beta_0\epsilon$

$$\beta_0\epsilon=\tau_y\left(\frac{8Re_3K^{2/n}}{\rho}\right)^{n/(n-2)}a^nD^{2n/(2-n)}, \quad (35)$$

with expressions for $\beta_1\epsilon^2$ and $\beta_2\epsilon^3$ found by setting the $\mathcal{O}[\epsilon]$ and $\mathcal{O}[\epsilon^2]$ terms to zero. Substituting $\beta_0\epsilon$ into the dominant term in (34) gives

$$V\sim\left(\frac{8^{n-1}KRe_3}{\rho D^n}\right)^{1/(2-n)}, \quad (36)$$

as the leading order term for the velocity in the small D limit. Note that $V\sim D^{-n/(2-n)}$ and so in the Bingham limit $n=1$, this implies that $V\sim 1/D$ as $D\to 0$. So, for small D the critical velocity (to leading order) is independent of the yield stress τ_y.

The condition $\beta_0\epsilon\ll 1$ indicates that the analysis is valid for

$$D\ll a^{-1+n/2}\left(\frac{8Re_3K^{2/n}}{\rho\tau_y^{(2-n)/n}}\right)^{1/2}. \quad (37)$$

As discussed in [9], an alternative way to derive (36) is to note that, since $r_p\ll R$ in the limit $D\to 0$, then $D_a\approx D$ and $V_a\approx V=\frac{D\tau_0^{1/n}a}{8K^{1/n}}$ to leading order in τ_y/τ_0. Then Re_3 in (4) becomes

$$Re_3\approx\frac{8\rho V^2}{K}\left(\frac{D}{8V}\right)^n, \quad (38)$$

where the τ_y term in the denominator has been neglected since $\tau_y\ll K\left(\frac{8V_a}{D_a}\right)^n$ for $\tau_0\gg\tau_y$. Rearranging (38) then gives (36) directly. Although this might seem like a more direct way to find an expression for the velocity, the analysis in terms of parameters $\mathcal{A}$ and $\mathcal{B}$ allows us to obtain the higher order corrections in a systematic way.

We now compare the large and small D asymptotic solutions with the exact solution and experimental data.

4. Results

The results are taken from a series of experiments described in [12, 16]: a number of fluids were pumped through a pipe test rig, over a range of flow rates, and the average velocity measured for a prescribed pressure gradient. The data points and fluid properties are provided in Table 1. Note that some obvious outliers in the data sets were omitted from our calculations. In laminar flow, the measured velocity agrees closely with that predicted by equation (16). When the flow becomes turbulent there is a marked departure between the measured and predicted values. This behaviour is depicted (in an idealised fashion) in Figure 2. The data points for V_c shown in Table 1 were extracted from pipe flow curves of the form shown in the figure using a normalised adherence function, see [17]. Note, transition phenomena, such as puffs, streaks etc.,

Data Set	ρ	τ_y	K	n	Data values (D_i, V_{ci})
A	1061.264	1.04	0.0136	0.8031	(0.005623, 1.458), (0.013214, 1.0725), (0.0216, 0.945)
B	1071.234	1.88	0.0102	0.8428	(0.005623, 1.4), (0.013214, 1.32), (0.0216, 1.134), (0.140512, 1.05), (0.207, 1.035)
C	1099.2	5.8	0.0176	0.8154	(0.079, 1.95), (0.140512, 1.875), (0.207, 1.87)
D	1105.034	4.18	0.0351	0.719	(0.005623, 2.11), (0.013214, 1.98), (0.0216, 1.62), (0.140512, 1.5152), (0.207, 1.55)
E	1176	3.91	0.0105	0.972	(0.0216, 1.935), (0.140512, 1.655), (0.207, 1.61)
F	1320.1	5.48	0.1239	0.6363	(0.0216, 1.79), (0.079, 1.74), (0.207, 1.553)
G	1394.35	8.02	0.135	0.5911	(0.079, 1.93), (0.140512, 1.94), (0.207, 1.87)
H	1501.276	8.0	0.03708	0.87907	(0.01337, 3.29), (0.02838, 2.33), (0.0042, 5.68)

Table 1. *Experimental data values [12, 16].*

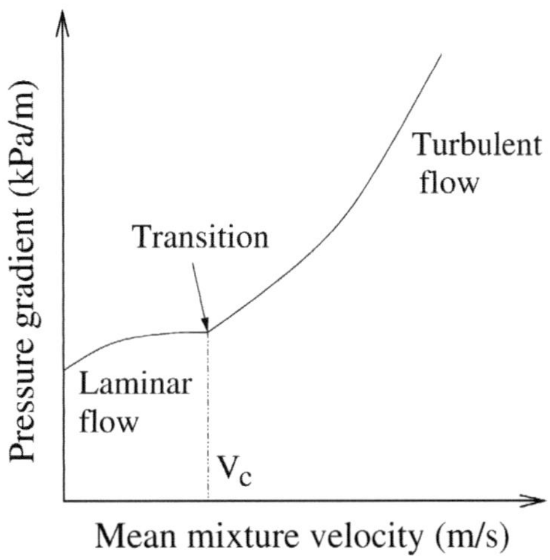

Figure 2. *Schematic of velocity against pressure gradient for water and slurry flows.*

are observed at values below V_c but these do not cause the pipe flow data to deviate from the laminar curve.

In the previous section we determined expressions for V_c in both the large and small D limits. To enable a comparison with the exact relation (20), with τ_0 found by solving the nonlinear expression in (19), we must obtain a value for Re_3 which gives the best fit to the experimental data. Equations (19) and (20) provide a relation between V (or V_c), Re_3 and D, where we must eliminate the unknown τ_0. One goal of this work is to determine the value for Re_3 that best predicts the transition. We calculate this using a nonlinear regression package provided on StatsPages.net [20]. For each given data point (D_i, V_i) we can determine a value for τ_{0i} using

IOP Conf. Series: Journal of Physics: Conf. Series **811** (2017) 012007 doi:10.1088/1742-6596/811/1/012007

(20). Then, re-writing (19) as

$$y = \frac{\tau_y + (\tau_0 - \tau_y)(\tau_0 + \tau_y)^{-n}[a(\tau_0 - \tau_y) + 2b\tau_y]^n}{(\tau_0 - \tau_y)^{2(n+1)/n}\tau_0^{-2}(\tau_0 + \tau_y)^{-2}[a(\tau_0 - \tau_y) + 2b\tau_y]^2} = \frac{\rho D^2}{8K^{2/n}Re_3},$$

we can determine a value for the unknown parameter Re_3 using the points (D_i, y_i). These values and the restrictions on D for the small and large diameter limits are given in Table 2 for each data set.

Data Set	Re_3	Limits (29), (37)
A	2135.5 ± 12.4	$D \gg 0.015$, $D \ll 0.019$
B	2148.5 ± 3.8	$D \gg 0.009$, $D \ll 0.012$
C	2402.7 ± 10.6	$D \gg 0.007$, $D \ll 0.008$
D	2330.0 ± 21.2	$D \gg 0.009$, $D \ll 0.012$
E	2554.6 ± 8.8	$D \gg 0.014$, $D \ll 0.019$
F	2109.6 ± 23.4	$D \gg 0.019$, $D \ll 0.024$
G	2361.9 ± 30.7	$D \gg 0.010$, $D \ll 0.012$
H	2374.0 ± 66.5	$D \gg 0.017$, $D \ll 0.023$

Table 2. *Best fit values of Re_3 for each data set.*

In Figures 3 and 4 we show plots of V_c against D for eight different Herschel-Bulkley fluids. The data points are denoted by a '*'. The solid line is the exact solution obtained by numerically solving equations (19) and (20). On the right hand side of each figure there are also three sets of curves. The lowest, represented by a dotted line, is the leading order solution for large D. It is clear from these plots that the critical velocity is independent of the pipe diameter in this limit. The solutions to first and second order are shown as dot-dash and dashed lines respectively. Similarly, on the left hand side are the three small D solutions. Finally, the two vertical lines represent the limit of validity for each expansion, which is given in the final column of Table 2. Note that these limits in fact overlap, in the sense that the small D limit is the right hand line and the large D limit is the left hand line. However, the criteria is that D is either much greater or much less than these limits and it is clear that, in most cases, the leading and first order solutions break down well before reaching these lines. Each figure requires a value for the critical Re_3, obtained by the nonlinear regression, these values are given in Table 2. Since the figures all take the same format we will only discuss the first, Figure 3 for Data Set A, in detail. In this case, the nonlinear regression, using 3 data points, indicates $Re_3 = 2135.5 \pm 12.4$. This small error is reflected in the fact that the exact solution, using $Re_3 = 2135.5$, appears to pass through all the data points. The leading order solution for large D provides an excellent approximation for pipe diameters around 1m (of the order 0.3%). Below this value the leading order solution predicts values less than the exact solution, with the accuracy decreasing with pipe diameter. The second order approximation appears accurate down to around 2cm (with an error around 5%). The restriction on D, given by equation (29), indicates that we require $D \gg 1.5$cm. For Data Set B we have more data points and find $Re_3 = 2148.5 \pm 3.8$. The two data points for large D sit on the exact solution curve, which at this stage is indistinguishable from the first and second order asymptotic expansions. The data point for the smallest value of D is somewhat below the exact solution curve. Although we do not have a great amount of data for small diameter pipes, it is clear from Data Set H, shown in Figure 4, that the exact solution does not in general over-predict V_c for small D.

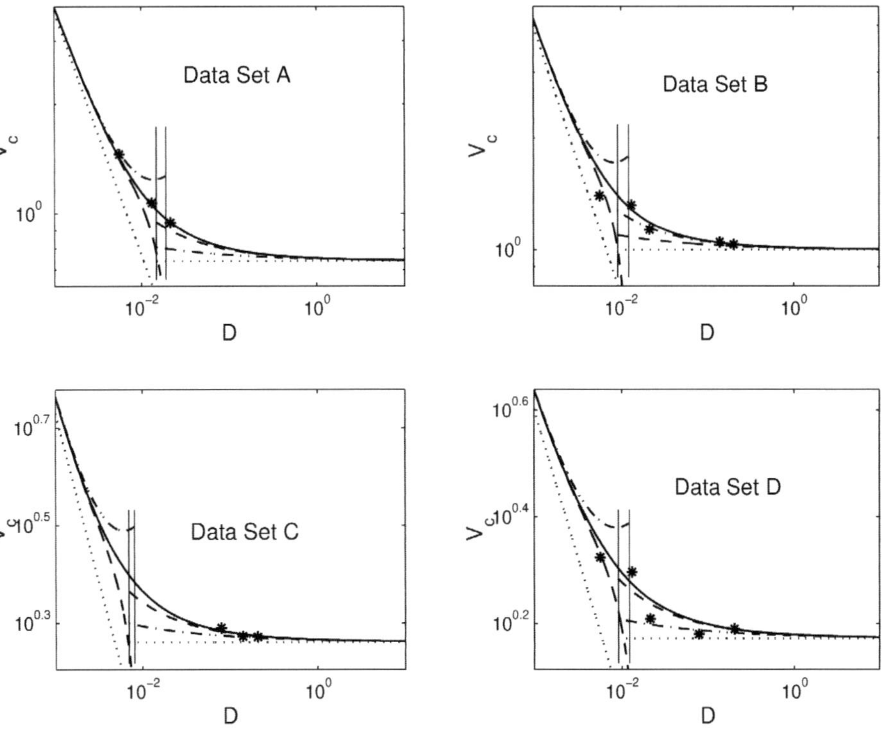

Figure 3. *Plot of critical velocity against pipe diameter for Data Sets A-D. The solid line denotes the exact solution, the dotted, dot-dashed and dashed lines denote the small (left) and large (right) D asymptotes for leading, first and second order corrections respectively.*

5. Conclusion and discussion

The primary aims of this paper were:

a) to determine whether Re_3 is a suitable expression for the non-dimensional Reynolds number which characterizes the flow of a yield-stress fluid and, if so, to find the best choice of Re_3 that predicts the transition to turbulence;

b) to investigate the asymptotic relation between the critical velocity and pipe diameter and, in particular, to determine the unknown coefficient $C(n)$ identified by dimensional analysis.

In general it appears that the form of Re_3 described by equations (19) and (20) leads to excellent agreement with the data for both large and small diameter pipes, and consequently it seems to be an appropriate non-dimensional grouping to characterize the flow. The nonlinear regression indicates that the transition to turbulence occurs for values of Re_3 within the range 2100 – 2500. This is in keeping with quoted values for a Newtonian fluid. There appears to be no obvious connection between the fluid properties and the predicted value of Re_3. For example, the three data sets that show $Re_3 \sim 2100$ have yield stresses ranging from 1.04 to 5.48 $\mathrm{N/m^2}$ and a power n between 0.84 and 0.64. The highest value of Re_3 predicted, $Re_3 \sim 2550$, occurred with the fluid most close to the Newtonian power, $n = 0.97$.

For fluids with an exponent $n \approx 1$, it is perhaps not surprising that the transition predicted by Re_3 is similar to Newtonian predictions. The thinking behind Re_3 is that only flowing material should be considered; the unyielded material is not included in the calculation. This is accounted for by the shift, of magnitude τ_y, in the denominator of Re_3. So, when $n = 1$,

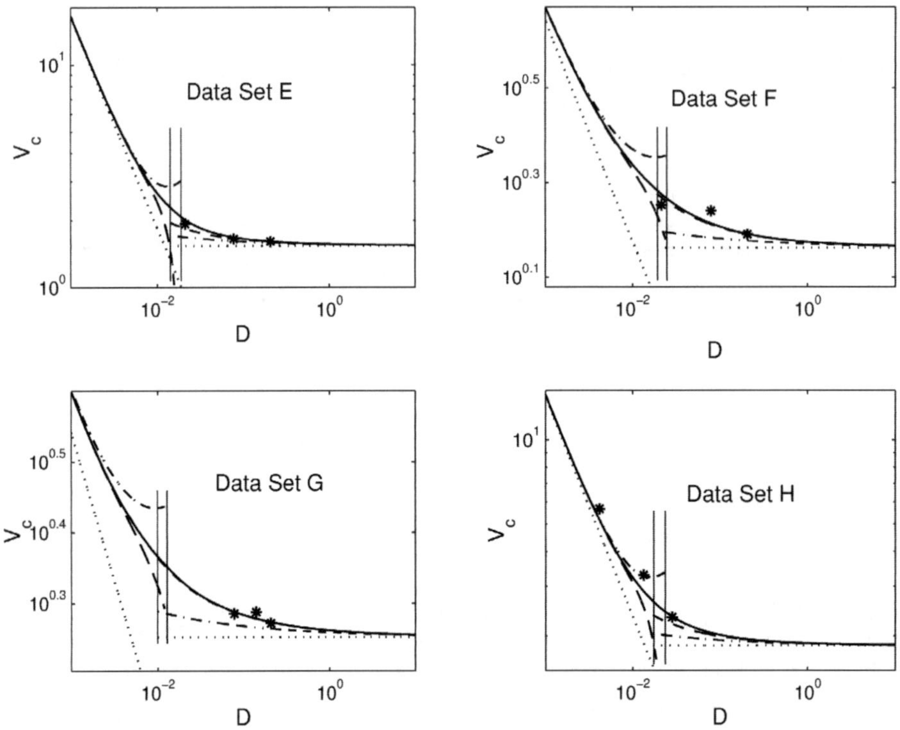

Figure 4. *Plot of critical velocity against pipe diameter for Data Sets E-H. The solid line denotes the exact solution, the dotted, dot-dashed and dashed lines denote the small (left) and large (right) D asymptotes for leading, first and second order corrections respectively.*

the flowing material exhibits Newtonian behaviour, the unyielded fluid is neglected and Re_3 is equivalent to the Newtonian Reynolds number (where V is replaced by V_a and D by D_a). Other definitions of the Reynolds number for a yield stress fluid take into account the unyielded material and so should not necessarily be expected to equate to the Newtonian value. For $n \neq 1$ the results of Draad *et al.* [2] suggest that with no yield stress the transition occurs for values of Re_w similar to the Newtonian case. Since $Re_3 \approx Re_w$ when $\tau_y = 0$ and $n \in [0.5, 1]$, by the same argument we should again expect transition to occur for $Re_3 \sim 2300$ when $\tau_y \neq 0$. Note, our experimental results satisfy the criteria since $n > 0.59$ for all cases; Draad *et al.* [2] have $n > 0.61$.

The actual calculation of Re_3 for given flow conditions is not as straightforward as for the Newtonian problem. It requires the calculation of the average annular velocity and the annular diameter which in turn require the average velocity and the wall stress. To determine a relation between V_c and D the wall stress must then be eliminated from expressions for Re_3 and V, given by equations (19) and (20). Clearly it is desirable to have a simple relation between the critical velocity and the pipe diameter, particularly for relatively large pipes which are of interest to a number of industries. Previously, dimensional analysis had shown that in the large pipe limit $V_c = C(n)\sqrt{\tau_y/\rho}$, where the coefficient $C(n)$ had to be calculated experimentally or numerically for a given fluid. The leading order term of the asymptotic analysis led to a simple form for $C(n)$, in terms of n and Re_3, and also indicated the lower limit for the pipe diameter when this could be applied. The first and second order corrections permitted this approximation to be extended to lower values of D. An asymptotic expansion for small diameter pipes showed that at leading order $V_c \propto D^{n/(n-2)}$.

Our work provides evidence confirming two physical facts of profound practical and theoretical importance for viscoplastic fluids:

- Firstly, at small pipe diameter, the critical velocity V_c is independent of the yield stress;
- Secondly, at large pipe diameter, V_c is independent of the pipe diameter and the consistency index, and the yield stress plays a dominant role.

Asymptotic relations describing both these conditions have been provided and tested against experimental data.

There is obvious scope for extending this work, to match the large and small diameter expansions to give a complete expression for V_c against D. We did not carry this out for two reasons. Firstly, our interest lays primarily with the large pipe limit, which is applicable to the transport of mining waste for example. Secondly, in our examples, the transition region was relatively small, typically between 1mm and 1cm, and the work involved seemed greater than simply calculating the exact solution numerically.

Acknowledgments
Financial support was provided through a Ministerio de Ciencia e Innovación grant MTM2014-56218 and the Mathematics Applications Consortium for Science and Industry (www.macsi.ul.ie) based at the University of Limerick, funded by the Science Foundation Ireland grant 12/IA/1683.

References

[1] R.P. Chhabra and J.F. Richardson, Non-Newtonian flow in the process industries, Butterworth-Heinemann, 1999.
[2] A.A. Draad and G.D.C. Kuiken and F.T.M. Nieuwstadt, Laminar-turbulent transition in pipe flow for Newtonian and non-Newtonian fluids, J. Fluid Mech., 1998, 377, 267-312.
[3] B. Eckhardt and T.M. Schneider, and B. Hof and J. Westerweel., Turbulence transition in pipe flow, Ann. Rev. Fluid Mech., 2007, 39, 447-468.
[4] M.P. Escudier and F. Presti, Pipe flow of a thixotropic liquid, J. Non-Newtonian Fluid Mech., 1996, 62, 291-306.
[5] G.W. Govier and K. Aziz, The Flow of Complex Mixtures in Pipes, Litton Educational Pub. Inc., 1972.
[6] H.C. Honey and W.A. Pretorius, Laminar flow pipe hydraulics of pseudoplastic-thixotropic sewage sludges, Water SA, 2000, 26(1), 19-25.
[7] R.R. Kerswell, Recent progress in understanding the transition to turbulence in a pipe, Nonlinearity, 2005, 18, 17-44.
[8] A.B. Metzner and J.C. Reed, Flow of non-Newtonian fluids - correlation of the laminar, transition and turbulent flow regions, AIChE J., 1955, 1(4).
[9] S.L. Mitchell & T.G. Myers, The laminar-turbulent transition of yield stress fluids in large pipes, Mathematics in Industry Study Group, 2007, January, University of Witwatersrand.
[10] J. Peixinho and C. Nouar and C. Desaubry and B.Théron, Laminar transitional and turbulent flow of yield stress fluid in a pipe, J. Non-Newtonian Fluid Mech., 2005, 128, 172-184.
[11] M. Rudman and H.M. Blackburn and L.J.W. Graham and L. Pullum, J. Non-Newtonian Fluid Mech., 2004, 118, 33-48.
[12] P.T. Slatter, The role of the yield stress on the laminar/turbulent transition, Ninth Int. Conf. on Transport & Sedimentation of Solid Particles, 1997, 547-561, Cracow, ISBN 8385582232.
[13] P.T. Slatter and E.J. Wasp, The laminar/turbulent transition in large pipes, Ninth Int. Conf. on Transport & Sedimentation of Solid Particles, 2001, 389-399, Wroclaw, September, ISBN 83-87866-12-1.
[14] P.T. Slatter and J.H. Lazarus, Critical flow in slurry pipelines, 12th Int. Conf. on slurry handling and pipeline transport, Hydrotransport 12, 1993, 639.
[15] P.T. Slatter, The role of rheology in the pipelining of mineral slurries, Min. Pro. Ext. Met. Rev., 1999, 20, 281-300.

[16] P.T. Slatter, The laminar/turbulent transition prediction for non-Newtonian slurries, Proc. Int. Conf. Problems in Fluid Mechanics and Hydrology", Academy of Sciences of the Czech Republic, 2001, Prague, June, 247-256, ISBN 80 238 3824 5.
[17] P.T. Slatter, A new friction factor for yield stress fluids, British Hydrodynamic Research Group, 14th Int. Conf. on slurry handling and pipeline transport, Hydrotransport 14, 1999, Maastricht, Sept., 255 - 265, ISBN 186058 213 3.
[18] R.J. Soto and V.L. Shah, Entrance flow of a yield-power law fluid, Appl. Sci. Res., 1976, 32, 73-85.
[19] F.T. Pinho and J.H. Whitelaw, Flow of non-Newtonian fluids in a pipe, J. Non-Newtonian Fluid Mech., 1990, 34, 129-144.
[20] StatsPages.net http://statspages.org/nonlin.html, last accessed Sept 23rd 2016.
[21] E.J. Wasp and J.P. Kenny and R.L. Gandhi, Solid-liquid flow: slurry pipeline transportation, Trans Tech Publications, 1977, New York.

Asymptotic analysis of reaction-diffusion-advection problems: Fronts with periodic motion and blow-up

Nikolay Nefedov
Department of Mathematics, Faculty of Physics, Lomonosov Moscow State University, 119899 Moscow, Russia
E-mail: nefedov@phys.msu.ru

Abstract. This is an extended variant of the paper presented at MURPHYS-HSFS 2016 conference in Barcelona. We discuss further development of the asymptotic method of differential inequalities to investigate existence and stability of sharp internal layers (fronts) for nonlinear singularly perturbed periodic parabolic problems and initial boundary value problems with blow-up of fronts for reaction-diffusion-advection equations. In particular, we consider periodic solutions with internal layer in the case of balanced reaction. For the initial boundary value problems we prove the existence of fronts and give their asymptotic approximation including the new case of blowing-up fronts. This case we illustrate by the generalised Burgers equation.

1. Introduction

We present some recent results of investigation of singularly perturbed reaction-advection-diffusion problems, which are based on a further development of the asymptotic comparison principle. We discuss further development of the general scheme of the asymptotic method of differential inequalities (see [1] - [5]) for periodic parabolic problems and apply this scheme to a few new cases. Theorems, which state the existence of periodic solutions with an internal layer, present an asymptotic approximation of such solution and establish Lyapunov stability for them. In particular, we consider periodic solutions with internal layer in the case of balanced reaction for the problems

$$\begin{gathered}
\varepsilon^2\left(\frac{\partial^2 u}{\partial x^2}-\frac{\partial u}{\partial t}\right)-\varepsilon^2 A(u,x,t)\,\frac{\partial u}{\partial x}-f(u,x,t,\varepsilon)=0 \quad \text{for} \quad x\in(0,1),\ t\in R,\\
\frac{\partial u}{\partial x}(0,t,\varepsilon)=u^{(-)}(t), \quad \frac{\partial u}{\partial x}(1,t,\varepsilon)=u^{(+)}(t) \quad \text{for} \quad t\in R,\\
u(x,t,\varepsilon)=u(x,t+T,\varepsilon) \quad \text{for} \quad t\in R,
\end{gathered}$$

where ε is a small parameter, A, f, $u^{(-)}(t)$ and $u^{(+)}(t)$ are sufficiently smooth T-periodic in t functions.

We also present an extension of these results for the reaction-advection-diffusion equation

$$\varepsilon\frac{\partial^2 u}{\partial x^2}-A(u,x)\frac{\partial u}{\partial x}-\frac{\partial u}{\partial t}=f(u,x,\varepsilon), \quad x\in(0,1),t>0.$$

For the initial boundary value problems for this equation we prove the existence of fronts and obtain their asymptotic approximation including the new case of the blowing-up fronts. This case is illustrated by the generalised Burgers equation. These results can be considered as an extension of the results of [7] - [9].

The paper is structured as follows. In Section 2 we consider the general scheme of our approach, extended to periodic parabolic problems. In Section 3 we apply it to a reaction-advection-diffusion equation with blowing-up fronts.

2. General scheme of asymptotic method of differential inequalities for periodic reaction-advection-diffusion equations

2.1. General scheme

We consider some cases of initial boundary value problem

$$Nu \equiv \varepsilon^2(\Delta u - \tfrac{\partial u}{\partial t}) - f(u, \nabla u, x, t, \varepsilon) = 0, \quad x \in \mathcal{D} \subset R^N,\, t \in R,$$
$$Bu = h(x,t), \quad x \in \partial\mathcal{D},\, t \in R,$$

with periodicity condition $u(x,t,\varepsilon) = u(x,t+T,\varepsilon)$, where ε is a small parameter, f, h, and $\partial\mathcal{D}$ are sufficiently smooth, B is a boundary operator for Dirichlet, Neumann or third type boundary conditions. We also assume that f, h are T-periodic in t functions.

Definition. *The T-periodic functions $\beta(x,t,\varepsilon)$ and $\alpha(x,t,\varepsilon)$ are called asymptotic upper and lower solutions of order $q > 0$ of the problem if for all sufficiently small ε they satisfy the inequalities*

$$N\beta \le -c\varepsilon^q,\ N\alpha \ge c\varepsilon^q, \quad x \in \mathcal{D},\, t \in R,$$
$$B\alpha \le h(x,t) \le B\beta, \quad x \in \partial\mathcal{D},\, t \in R,$$

where c is a positive constant.

We start with the description of our approach to investigate the existence and Lyapunov stability of periodic solutions with boundary and internal layers.

Denote by L the linear operator which we get from N by linearizing f on the periodic solution, by L_f the linearization f on the periodic solution and by H the following characteristic of the nonlinearity

$$H \equiv f(\beta, \nabla\beta, x, t, \varepsilon) - f(\alpha, \nabla\alpha, x, t, \varepsilon) - L_f(\beta - \alpha).$$

The properties of H depend on the asymptotic lower and upper solutions. In what follows we describe the method to construct them by using the formal asymptotic expansion. We assume that

(A_1). *There exist an asymptotic upper solution β and an asymptotetic lower solution α of order q such that $\beta > \alpha$ and $|\beta - \alpha| \le c\varepsilon^r$ for $x \in \mathcal{D},\, t \in R$ and all sufficiently small ε.*

We also assume that

(A_2). $|H| \le c\varepsilon^p$ for $x \in \mathcal{D},\, t \in R$ and a sufficiently small ε.

(A_3). $p \ge q$.

Theorem 1. *Suppose that assumptions $(A_1) - (A_3)$ are valid. Then, for any sufficiently small ε there exists a solution $u(x,t,\varepsilon)$ which differs from the upper and lower solutions by the value of order $O(\varepsilon^r)$ and is asymptotically stable in Lyapunov sense with the local domain of stability $[\alpha, \beta]$.*

The existence of the solution follows from assumption (A_1) (see [10]) . The proof of stability is based on a revised maximum principal, which uses Krein-Rutman theorem. The scheme of the proof can be found in [1] (see also [4],[11] and references therein).

In order to obtain the upper and lower solutions satisfying the assumptions of Theorem 1 we use the formal asymptotic expansion, which can be constructed in a lot of cases by our method. Under quite natural assumptions the formal asymptotic expansion of internal layer solution is produced by the boundary layer operators L_B, regular expansion operators L_R and by the operators of order q describing the location of the transition layer. To construct the formal asymptotic expansion we assume that the operators are invertible.

For the construction of asymptotic lower and upper solutions we require that **these operators are monotone (order preserving) when they act in the same classes of functions in which we construct the asymptotic expansions by means of these operators.**

Finally we obtain $\alpha \equiv \alpha_n$, $\beta \equiv \beta_n$, the modified n-th order formal asymptotic expansions.

2.2. *Periodic BVP with Balanced Reaction*

We consider periodic solutions with internal layer in the case of the balanced reaction for the problem

$$\begin{aligned} \varepsilon^2\left(\frac{\partial^2 u}{\partial x^2} - \frac{\partial u}{\partial t}\right) - \varepsilon^2 A(u,x,t)\,\frac{\partial u}{\partial x} - F(u,x,t,\varepsilon) = 0 \quad \text{for} \quad x \in (0,1),\ t \in R, \\ \frac{\partial u}{\partial x}(0,t,\varepsilon) = u^{(0)}(t), \quad \frac{\partial u}{\partial x}(1,t,\varepsilon) = u^{(1)}(t), \quad t \in R, \\ u(x,t,\varepsilon) = u(x,t+T,\varepsilon) \quad \text{for} \quad t \in R, \end{aligned} \tag{1}$$

where ε is a small parameter, A, F, $u^{(0)}(t)$ and $u^{(1)}(t)$ are sufficiently smooth T-periodic in t functions.

We assume, that

- (A_1). The reduced equation $F(u,x,t,0) = 0$ has exactly three solutions $\phi^{(-)}(x,t), \phi^{(0)}(x,t), \phi^{(+)}(x,t)$, satisfying
$$\phi^{(-)}(x,t) < \phi^{(0)}(x,t) < \phi^{(+)}(x,t);$$
- (A_2). $F_u(\phi^{(\pm)},x,t,0) > 0, F_u(\phi^{(0)},x,t,0) < 0$;

and consider the case of the balanced reaction

- (A_3). $I(x,t) := \int_{\phi^{(-)}(x,t)}^{\phi^{(+)}(x,t)} F(u,x,t,0)du \equiv 0$.

To construct the formal interior layer asymptotic expansion we use the ansatz (a similar treatment one can find in [1, 2, 3, 4, 5])

$$U^{(\pm)}(x,t,\varepsilon,x^*) = \overline{u}^{(\pm)}(x,t,\varepsilon) + Q^{(\pm)}(\tau,x^*,t,\varepsilon) + \Pi^{(\pm)}(\xi,t,),$$

where the terms are the series in ε and the location of the interior layer $x^*(t,\varepsilon)$ is *a priori* unknown. This location is defined by the C^1-matching procedure applied to $U^{(\pm)}(x,t,\varepsilon,x^*)$ with $x^*(t,\varepsilon) = x_0(t) + \varepsilon x_1(t) + \varepsilon^2 x_2(t) + \ldots$, $\xi = x/\varepsilon$, $x \le x^*$, $\xi = (x-1)/\varepsilon$, $x \ge x^*$, and $\tau = (x - x^*)/\varepsilon$ (symbols "-" and "+" are used for the terms of asymptotic expansions in the domains on the left and right sides of the curve $x^*(t)$, respectvely). The terms of the asymptotic expansions are defined in a standard way, see for example [1, 2, 3, 4, 5].

The main term of the layer location is defined by the problem

$$-\frac{dx_0}{dt}\int_{-\infty}^{+\infty} \Phi^2(\tau,x_0,t)d\tau + \int_{-\infty}^{+\infty}\left(\Phi^2(\tau,x_0,t)\tilde{A} + \Phi(\tau,x_0,t)\tilde{F}_x\tau + \Phi(\tau,x_0,t)\tilde{F}_\varepsilon\right)d\tau = 0$$

$$x_0(t) = x_0(t+T),$$

where " $\sim$ " means that the function is evaluated at the point $(\tilde{u}^{(\pm)}(\tau,x_0,t),x_0,t,0)$, and $\Phi^{(\pm)}(\tau,x_0,t) = \frac{\partial Q_0^{(\pm)}}{\partial \tau}(\tau,x_0,t)$.

We assume that

- (A_4). The problem for $x_0(t)$ has a solution.

For the next terms of the layer location we have

$$m(t)\frac{dx_i}{dt} - d(t)x_i = f_i(t), \quad i = 1, 2, ..., \quad x_i(t) = x_i(t+T),$$

where, at each step, the functions $f_i(t)$ are known and

$$m(t) := \int_{-\infty}^{+\infty} \Phi^2(\tau, x_0, t)d\tau,$$

$$d(t) := -\frac{dx_0}{dt}\frac{\partial\left(\int_{-\infty}^{+\infty} \Phi^2(\tau,x_0,t)d\tau\right)}{\partial x_0} + \frac{\partial\left(\int_{-\infty}^{+\infty}\left(\Phi^2(\tau,x_0,t)\tilde{A} + \Phi(\tau,x_0,t)\tilde{F}_x\tau + \Phi(\tau,x_0,t)\tilde{F}_\varepsilon\right)d\tau\right)}{\partial x_0}.$$

A further assumption

- (A_5). $\frac{1}{T}\int_0^T \frac{d(t)}{m(t)}dt < 0.$

guarantees that the problem for $x_i(t)$ is solvable and the first order periodic operator is an order preserving operator.

According to the previous section we construct an upper solution in the form (similar construction one can find, for example, in [4])

$$\beta_n(x,t,\varepsilon) = \bar{u}_0^{(\pm)}(x,t) + \varepsilon\bar{u}_1^{(\pm)}(x,t) + \ldots + \varepsilon^{n+3}\bar{u}_{n+3}^{(\pm)}(x,t)$$
$$+Q_0^{(\pm)}(\tau_\beta, x_{n\beta}, t) + \varepsilon Q_1^{(\pm)}(\tau_\beta, x_{n\beta}, t) + \ldots + \varepsilon^{n+2}Q_{n+2}^{(\pm)}(\tau_\beta, x_{n\beta}, t)$$
$$+\Pi_\beta(\xi, t, \varepsilon) + \varepsilon^{n+3}(\gamma + Q_{n+3,\beta}^{(\pm)}(\tau_\beta, x_{n\beta}, t)),$$

where $x_{n\beta}(t,\varepsilon) = x_n^*(t,\varepsilon) + \varepsilon^{n+2}(x_{n+2}(t) - \nu(t))$, $\gamma > 0$ is a positive constant and $\nu(t)$ is a positive solution of the the problem in (A_5) with a positive right hand side.

The lower solution is constructed analogously.

We can follow the proof for the unbalanced case of problem (1) considered in [4] to check that $\beta_n(x,t,\varepsilon)$ and $\alpha_n(x,t,\varepsilon)$ are asymptotic upper and lower solutions and that q and p in Theorem 1 are $q = n+3$, $p = 2n+2$. Therefore the assumption $p \geq q$ is valid for $n \geq 1$. We have

Theorem 2 *Suppose* $(A_1)-(A_5)$ *are valid. Then there exists a contrast structure type solution* $u(x,t,\varepsilon)$ *of problem (1):*

$$|u(x,t,\varepsilon) - U_n(x,t,\varepsilon)| < C\varepsilon^{n+1},$$

which is asymptotically stable in the sense of Lyapunov with the local domain of stability $[\alpha_1, \beta_1]$, *where* $U_n(x,t,\varepsilon)$ *is the n-th order partial sum of the constructed approximation and* α_1, β_1 *are lower and upper solutions constructed by the modification of* $U_4(x,t,\varepsilon)$.

3. Initial Boundary Value Problems with Fronts: Motion and Blow-Up

We present recent results for some classes of IBVP (initial boundary value problem) where we investigate moving fronts by using the developed comparison technique. For these initial boundary value problems we proved the existence of fronts and give its asymptotic approximation. We proved that the principal term, describing the location of the moving front, is determined by the initial value problem (see [7])

$$\frac{dx_0}{dt} = V(x_0), \quad x_0(0) = x_{00}, \tag{2}$$

where x_{00} is the initial location of the front, $V(x_0)$ is a known function, defined by the input data. We proved that the Lyapunov stability of steady points of equation (2) determine the Lyapunov stability of stationary solutions with interior layer of the IBVP. In the present paper we also have proved that under some conditions the blow-up of the solution problem (2) determine the blow-up of the interior layer solution of the IBVP.

3.1. *Reaction-advection-diffusion equations*

We illustrate our results by the problem

$$\varepsilon\frac{\partial^2 u}{\partial x^2} - A(u,x)\frac{\partial u}{\partial x} - \frac{\partial u}{\partial t} = f(u,x,\varepsilon), \quad x \in (0,1), t > 0,$$

$$u(0,t,\varepsilon) = u^0, \quad u(1,t,\varepsilon) = u^1, \quad t \in [0,T],$$
$$u(x,0,\varepsilon) = u_{init}(x,\varepsilon), \quad x \in [0,1].$$

As it was mentioned above $u_{init}(x,\varepsilon)$ is a front type function which is approximated by the formal asymptotics and therefore the initial location x_{00} is considered as the principal term of the location expansion. The generation of the front is not considered.

We assume that

- (H_1). The equation
$$A(u,x)\frac{du}{dx} + B(u,x) = 0$$
with the initial condition $u(0) = u^0$ has a solution $u = \varphi^l(x)$, and the same equation with the initial condition $u(1) = u^1$ has a solution $u = \varphi^r(x)$. Moreover,
$$\varphi^l(x) < \varphi^r(x), \quad x \in [0,1]$$
and
$$A(\varphi^l(x),x) > 0, \quad A(\varphi^r(x),x) < 0, \quad x \in [0,1].$$
- (H_2).
$$I(x) := \int_{\varphi^l(x)}^{\varphi^r(x)} A(u,x)du > 0.$$
- (H_3). The initial value problem
$$\frac{dx_0}{dt} = \frac{I(x_0)}{\varphi^r(x_0) - \varphi^l(x_0)} \equiv V(x_0), \quad x_0(0) = x_{00},$$
where x_{00} is the initial location of the front, has a solution $x_0(t) : [0;T] \to [0,1]$ such that
$$x_0(t) \in (0,1) \quad \text{for} \quad t \in [0,T].$$

The main result for this problem is a theorem on the existence and asymptotic approximation of the moving front with the principal term $x_0(t)$ of the front location. The main result of [7] is that asymptotically stable stationary points of the equation of motion

$$\frac{dx_0}{dt} = V(x_0)$$

produce stable stationary solutions with the fronts of the original parabolic problem located near these points.

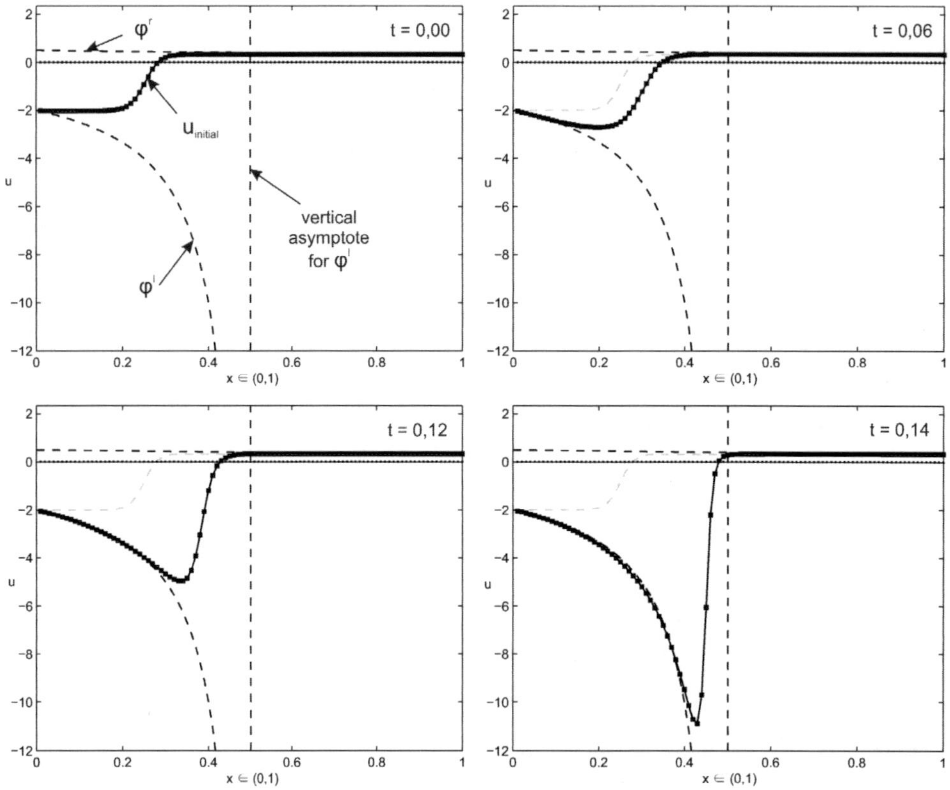

Figure 1. A typical example of a solution $u(x,t)$ to the Burgers type equation (a refinement of the mesh in a neighborhood of the transition point and the bounds has been performed).

3.2. Burgers type equation with blow-up of the solution

Let us replace assumption (H_1) with the following assumption:

- (H_1'). The equation
$$A(u,x)\frac{du}{dx} + B(u,x) = 0$$
with the initial condition $u(0) = u^0$ has a solution $u = \varphi^l(x)$ with blow-up near some point x_c, i.e. $\varphi^l(x) \to -\infty$ for $x \to x_c$, and the same equation with the initial condition $u(1) = u^1$ has a solution $u = \varphi^r(x)$. Moreover,
$$\varphi^l(x) < \varphi^r(x), \quad x \in [0, x_c)$$
and
$$A(\varphi^l(x), x) > 0, \quad x \in [0, x_c) \quad A(\varphi^r(x), x) < 0, \quad x \in [0, 1].$$

We have shown that the Burgers type equation can exhibit a blow-up of the front type solution with the jump at the front tending to infinity. We illustrate this phenomenon by the following

Burgers type equation with cubic forcing:

$$\varepsilon\frac{\partial^2 u}{\partial x^2} - \frac{\partial u}{\partial t} = -u\frac{\partial u}{\partial x} - u^3, \quad x \in (0,1), \quad t \in (0, 0.3],$$
$$u(0,t) = -2, \quad u(1,t) = \frac{1}{3},$$
$$u(x,0) = \frac{7}{6}\tanh\frac{x-\frac{1}{4}}{\varepsilon} - \frac{5}{6}.$$

An example of calculations for this problem is shown in Figure 1.

Acknowledgments

The author thanks the referees for important comments that helped to improve this paper. This work is supported by RFBR, pr. N 16-01-00437, and RFBR-DFG, pr. N 14-01-91333.

References

[1] Nefedov, N. 2013 *Comparison Principle for Reaction-Diffusion-Advection Problems with Boundary and Internal Layers.* Lecture Notes in Computer Science 8236, 62 - 72.
[2] Vasileva, A.B. and Butuzov, V.F. and Nefedov N.N. 2010. *Singularly Perturbed problems with Boundary and Internal Layers.* Proceedings of the Steklov Institute of Mathmatics, Vol.268, 258 - 273.
[3] Nefedov, N.N. and Recke, L. and Schneider K.R. 2013. *Existence and asymptotic stability of periodic solutions with an interior layer of reaction-advection-diffusion equations.* Journal of Mathematical Analysis and Applications 405, 90 - 103.
[4] Nefedov, N. N. and Nikulin, E. I. 2015. *Existence and stability of periodic contrast structures in the reaction-advection-diffusion problem.* Russian Journal of Mathematical Physics. Vol. 22, no. 2, 215 - 226.
[5] Butuzov, V. F. and Nefedov, N. N. and Recke, L. and K. R. Schneider. 2014. *Periodic solutions with a boundary layer of reaction-diffusion equations with singularly perturbed neumann boundary conditions*, International Journal of Bifurcation and Chaos, 24, 1440019 - 1 - 14400198.
[6] Vasil'eva, A.B. and Butuzov, V.F. and Nefedov, N.N. 2010. *Singularly Perturbed problems with Boundary and Internal Layers.* Proceedings of the Steklov Institute of Mathmatics, 268, 258-273.
[7] Nefedov, N. N. 2016. *Multiple scale reaction-diffusion-advection problems with moving fronts.* Journal of Physics: Conference Series, 727, 012011.
[8] Nefedov, N.N. and Nikitin, A.G. and Petrova, M.A. and Recke, L. 2011. *Moving fronts in integro-parabolic reaction-advection-diffusion equations.* Diff. Equations 47, 1318-1332.
[9] Bozhevol'nov, Yu.V. and Nefedov N.N. 2010. *Front motion in parabolic reaction-diffusion problem.* Computational Mathematics and Mathematical Physics 50(2), 276-285.
[10] Hess, P. 1991 *Periodic-Parabolic Boundary Value Problems and Positivity.* Pitman Research Notes in Math. Series 247, Longman Scientific & Technical, Harlow.
[11] Nefedov, N.N. and Recke, L. and Schneider, K.R. 2010. *Asymptotic stability via the Krein-Rutman theorem for singularly perturbed parabolic periodic Dirichlet problems.* Regular and Chaotic Dynamics, 15(2-3), 382-389.

Regularization of the matrix Riccati equation in optimal estimation problem with low measurement noise

Mikhail Osintcev[1] **and Vladimir Sobolev**[2]

[1] Department of Mathematics, North Carolina State University, Raleigh, NC 27695, USA
[2] Samara National Research University, 34, Moskovskoe shosse, Samara, 443086, Russian Federation

E-mail: [1] mishaosintsev@gmail.com, [2] hsablem@yahoo.com

Abstract. We explore the possibility of applying the method of order reduction of optimal estimation problem for singularly perturbed systems with low measurement noise. It is shown that matrix Riccati equation for the Kalman-Bucy filter has a periodic solution. An optimal filter is constructed for a dynamic model of a crank mechanism.

1. Introduction and statement of problem

Consider the optimal filtering problem for linear system:

$$\dot{x}(t) = A(t,\mu)x(t) + B(t,\mu)w(t), \tag{1}$$

where $x(t,\mu) \in R^n$ is the system state vector, $t \in R$ is the time, $\mu \ll 1$ is a small positive parameter, $A(t,\mu) \in R^{n\times n}$ is the dynamic coefficient matrix, $B(t,\mu) \in R^{n\times m}$ is the process noise coupling matrix, $w \in R^m$ is the zero-mean white Gaussian process noise with given covariance $Q \in R^{m\times m}$. We assume that the following quantity can be measured:

$$z(t) = C(t,\mu)x(t) + \mu v(t), \tag{2}$$

where $z(t) \in R^k$ is the measurement vector, $C(t,\mu) \in R^{k\times n}$ is the measurement sensitivity matrix, $v \in R^k$ is the zero-mean white Gaussian measurement noise with covariance $R(t,\mu) \in R^{k\times k}$. Let $w(t)$ and $v(t)$ be independent. The presence of the small parameter in (2) means that optimal estimation problem with low measurement noise (see [1] and references therein) is considered.

We are required to obtain an estimate $\hat{x}(t)$ of the state $x(t)$ of system (1) in accordance with the vector function $z(t)$ available for measurement at $t > 0$. The solution to this problem may be obtained by using the Kalman-Bucy filter [2] that involves the solution of the matrix Riccati equation for the covariance matrix of the filter [3]:

$$\mu^2(\dot{P} - AP - PA^T - BQB^T) = -PC^TR^{-1}CP, \quad P(0) = P_0. \tag{3}$$

The asymptotic expansion of solution to the Riccati equation for time-invariant matrices A, B, C, Q, R in the case of optimal control problem was obtained in [4].

The Riccati equation (3) may be considered as a singularly perturbed ODE system [5, 6] in a critical case since the corresponding limiting problem (under $\mu = 0$)

$$PC^T R^{-1} CP = 0 \tag{4}$$

possesses multiple zero solutions [7, 8].

The solution to the differential equation (3) may be separated into the following cases for $t > 0$:

Case 0: matrix BQB^T is positive definite and $rank(C) = n$;

Case 1: matrix $CBQB^T C^T$ is non-singular and $rank(C) = r \leq n$;

Case L $(L \geq 2)$ *:*

$$C_j BQB^T C_j^T = 0, \quad j = \overline{0,\, L-2}, \quad C_{L-1} BQB^T C_{L-1}^T > 0, \tag{5}$$

where $C_0 = C$, $C_j = C_{j-1}A^T - \dot{C}_{j-1}$. The solution to the problem (3) in cases 0 and 1 may be obtained as asymptotic expansion in integer powers of small parameter μ. In case L one may obtain solution of the problem (3) as asymptotic expansion in fractional powers of small parameter $\varepsilon = \mu^{1/L}$ [4]. Thus, neglecting by the initial value condition, we will try to find the partial solution in the following form :

$$P(t, \mu) = \sum_{j=0}^{\infty} P_j(t) \mu^{j/L}.$$

In general, solutions of this kind are the solutions of differential subsystem describing the flow of the original system on the slow integral manifold [7, 9, 10]. The integral manifolds method was used for the investigation of optimal control and filtering problems in [11, 12, 13, 14, 15].

In the paper we use a combination of geometric and asymptotic approaches for the regularization of matrix Riccati equations in some critical cases. Firstly, we try to find a formal solution to the matrix Riccati equation as asymptotic expansions

$$p_k = \sum_{j=0}^{\infty} p_{kj}(t) \varepsilon^j.$$

Here, p_k is some element of matrix P and $\varepsilon = \mu^{1/L}$. Let p_{ks_k} be the first nonzero coefficient in the expansion to p_k, then we introduce a new variable $p_k = \varepsilon^{s_k} y_k$. The corresponding ODE system for new variables possesses the nonzero solution $p_k = p_k(t, \varepsilon)$, $p_k(t, 0) \neq 0$ and we can obtain singularly perturbed ODE system in a non-critical case.

2. Simple example

Consider the optimal estimation problem for the following system:

$$\ddot{x}(t) + ax(t) = w(t), \quad x(0) = x_0, \tag{6}$$

where $x(t)$ is a scalar function, a is a constant coefficient, $w(t)$ is a scalar zero-mean white Gaussian process noise with given dispersion $q = 1 + 2\mu$, where $\mu \ll 1$ is a small positive parameter. The following function is available for measurement:

$$z(t) = x(t) + \mu v(t), \tag{7}$$

where v is a zero-mean scalar white Gaussian measurement noise with given dispersion $r = 1$. Functions $w(t)$ and $v(t)$ are independent. Obviously, this problem is similar to the problem (1) - (2). In this case, we have

$$A = \begin{pmatrix} 0 & 1 \\ -a & 0 \end{pmatrix}, B = \begin{pmatrix} 0 \\ 1 \end{pmatrix}, R = (\mu^2), Q = (q),$$

$$C = \begin{pmatrix} 1 & 0 \end{pmatrix}, P = \begin{pmatrix} p_0 & p_1 \\ p_1 & p_2 \end{pmatrix}.$$

It is easy to see that in the case under consideration, $L = 2$.

We use $a = 1$ for simplicity and construct the matrix Riccati equation (3) for covariance matrix of the Kalman-Bucy filter:

$$\begin{aligned} \dot{p}_0 &= 2p_1 - \frac{p_0^2}{\mu^2}, \\ \dot{p}_1 &= p_2 - p_0 - \frac{p_0 p_1}{\mu^2}, \\ \dot{p}_2 &= 1 + 2\mu - 2p_1 - \frac{p_1^2}{\mu^2}. \end{aligned} \tag{8}$$

The reduced system (4) has the following solution:

$$p_0 = \mu\sqrt{2\mu}, \quad p_1 = \mu, \quad p_2 = \sqrt{2\mu}(1+\mu). \tag{9}$$

System (6) may be integrated explicitly and has the following solution:

$$p_0 = \mu\sqrt{2\mu}\Bigg(1 - \frac{2e^{-\sqrt{\frac{2}{\mu}}t}}{D(t)}\Bigg(\nu^2\cos\left(\sqrt{\frac{2}{\mu}}\frac{t}{\nu}\right) + \nu\sin\left(\sqrt{\frac{2}{\mu}}\frac{t}{\nu}\right) + 1 + \nu^2 + e^{-\sqrt{\frac{2}{\mu}}t}\Bigg)\Bigg),$$

$$p_1 = \mu\Bigg(1 - \frac{2e^{-\sqrt{\frac{2}{\mu}}t}}{D(t)}\Bigg((\nu^2+1)\cos\left(\sqrt{\frac{2}{\mu}}\frac{t}{\nu}\right) + 1 + \nu^2\Bigg)\Bigg),$$

$$\begin{aligned} p_2 = \sqrt{2\mu}(1+\mu)\Bigg(1 + \frac{2e^{-\sqrt{\frac{2}{\mu}}t}}{D(t)}\Bigg(&-\nu^2\cos\left(\sqrt{\frac{2}{\mu}}\frac{t}{\nu}\right) \\ &+ \left(\mu + \frac{\nu}{(1+\mu)}\right)\sin\left(\sqrt{\frac{2}{\mu}}\frac{t}{\nu}\right) - 1 - \nu^2 - e^{-\sqrt{\frac{2}{\mu}}t}\Bigg)\Bigg), \end{aligned}$$

where $\nu = \sqrt{1/(1+2\mu)}$ and

$$D(t) = 1 + e^{-2\sqrt{\frac{2}{\mu}}t} + e^{-\sqrt{\frac{2}{\mu}}t}\left(\frac{1}{1+\mu}(\nu^2+1)\cos\left(\sqrt{\frac{2}{\mu}}\frac{t}{\nu}\right) + 2(1+\nu^2)\right).$$

We conclude that the solution of the system (8) tends exponentially to the steady state. Moreover, the difference between two this solutions is of order $O(e^{-\sqrt{\frac{2}{\mu}}t})$ as $\mu \to 0$, $t > 0$. It means that with very high order of accuracy the steady state may be used instead the exact solution.

IOP Conf. Series: Journal of Physics: Conf. Series **811** (2017) 012009 doi:10.1088/1742-6596/811/1/012009

3. Optimal estimation problem for parametric oscillator

Consider the model of parametric oscillator without damping, but with the presence of process noise on the right side of equation:

$$\ddot{x} + a(t)x = w, \tag{10}$$

where $a(t)$ is a continuous and uniformly bounded together with a sufficient number of derivatives function, w is a zero-mean white Gaussian process noise with a given correlation q. We assume that the following quantity can be measured:

$$z = x + \mu v, \tag{11}$$

where v is a zero-mean white Gaussian measurement noise with a given correlation r, and $\mu \ll 1$ is a small positive parameter. Problem (10)-(11) naturally appears when considering the optimal estimation problem in mechanics. For example, a simple crank mechanism is an arm connected to a rotating shaft by which reciprocating motion is imparted to the shaft [16]. The dynamic model of this system is

$$\ddot{\phi} + c\left(\frac{1}{I_m} + \frac{1}{I_0}\right)\left(1 - \frac{I_M m_2 r^2}{2I_0(I_M + I_0)}\cos 2\pi t\right)\phi = w, \tag{12}$$

where ϕ is the angle of the shaft, I_0 is the average moment of inertia of the crank, I_M is the moment of inertia of the shaft, m_2 is the arm mass and c is the stiffness coefficient of the arm. The system is influenced by an external white Gaussian disturbance w.

The solution of the optimal estimation problem (10)-(11) may be obtained by Kalman-Bucy filter that involves solution of matrix Riccati equation of the following form:

$$\begin{aligned} \mu^2(\dot{p}_0 - 2p_1) &= -\frac{p_0^2}{r}, \\ \mu^2(\dot{p}_1 + a(t)p_0 - p_2) &= -\frac{p_0 p_1}{r}, \\ \mu^2(\dot{p}_2 + 2a(t)p_1 - q) &= -\frac{p_1^2}{r}. \end{aligned} \tag{13}$$

We note that numerical solution of the singularly perturbed system is the most expensive stage in the construction of the filter. One should use sufficiently small time step in the numerical simulations to achieve the desired accuracy of the filter. This issue is of significant importance in real-time algorithms with limited computational resources.

We introduce variables p_i, $i = 0..2$:

$$p_0 = \varepsilon^3 y_0, \quad p_1 = \varepsilon^2 y_1, \quad p_2 = \varepsilon y_2, \tag{14}$$

where $\varepsilon^2 = \mu$. Thus, system of equations (13) takes the following form:

$$\begin{aligned} \varepsilon\dot{y}_0 &= 2y_1 - \frac{y_0^2}{r}, \\ \varepsilon\dot{y}_1 &= y_2 - \varepsilon^2 a(t)y_0 - \frac{y_0 y_1}{r}, \\ \varepsilon\dot{y}_2 &= q - 2\varepsilon^2 a(t)y_1 - \frac{y_1^2}{r}. \end{aligned} \tag{15}$$

This singularly perturbed system is not in critical case and usually the integral manifolds method is used in such a situation [15]. It is possible to construct attracting periodic solution for this system. We construct the periodic solution for system (15) with accuracy $O(\varepsilon^3)$. This means that the Riccati matrix equation is replaced by the periodic solution which with accuracy $O(\mu^6)$ has the form

$$\begin{aligned} p_0 &= \mu^3 \sqrt[4]{4qr^3} - \mu^5 \frac{a(t)r^2}{\sqrt[4]{4qr^3}} + \mu^6 \frac{\dot{a}(t)r^2}{2\sqrt{qr}}, \\ p_1 &= \mu^2 \sqrt{qr} - \mu^4 a(t) r + \mu^5 \frac{\dot{a}(t)r}{2\sqrt{2}} \sqrt[4]{\frac{r}{q}}, \\ p_2 &= \mu \sqrt[4]{4q^3 r} - \mu^3 \frac{a(t)\sqrt[4]{qr^3}}{\sqrt{2}}. \end{aligned} \tag{16}$$

4. Numerical experiments

Figure 2 shows comparison of the solution of system (13) for Kalman-Bucy filter covariance for the crank mechanism model with the periodic solution (16). When using the latter, we don't need to solve any system of differential equations because the components of the covariance matrix for the filter may be computed explicitly. The following parameters are used: $c = 1$, $I_0 = 50$, $I_m = 1$, $m_2 = 1.2$, $r = 1$, $\mu = 0.05$, $a(t) = \cos(t)$.The RK4 method is used for solving system (13) with time step $dt = 1 \times 10^{-4}$. We see that the solution of the original system tends to the periodic solution exponentially, beyond that, the difference between the solutions for p_0 is of the order 1×10^{-10}, for p_1 is 1×10^{-9} and for p_2 is 1×10^{-7} (see Figure 3).

5. Conclusion

In this paper, we considered the optimal estimation problem with low measurement noise. The matrix Riccati system of differential equations for the Kalman-Bucy filter is singularly perturbed. The method of model order reduction is applicable for this differential system even if the system is in critical case. It was shown that the periodic solution may be used instead the exact initially value problem for the matrix Riccati equation for covariance matrix of the filter. This approach is demonstrated on model of crank mechanism. Our numerical experiment shows that the difference between the solution of the original matrix Riccati differential system and the periodic solution is of the order $O(\mu^6)$. The evaluated example of crank mechanism does not account for dumping, Coulomb type friction, and structure elasticity, but is used as an example to show the approach to the reduction the order of optimal estimation problems with low measurement noise.

Acknowledgments

The reported study was funded by RFBR and Samara Region according to the research project 16-41-630524 and the Ministry of education and science of the Russian Federation in the framework of the implementation of Program of increasing the competitiveness of SSAU for 2013–2020 years.

References

[1] Gajic Z and Myo-Taeg L 2001 *Optimal control of singularly perturbed linear systems and applications* (New-York: CRC Press) p 326

[2] Kalman R E, Bucy R S 1961 *New results in linear filtering and prediction theory* (Trans. ASME, Ser. D, J. Basic Eng) p 109

[3] Anderson B D O, Moore J B 1989 *Optimal Control: Linear Quadratic Methods* (Prentice-Hall) p 394

[4] O'Malley R E, Jameson A 1975-77 Singular perturbations and singular arcs I, II. *Trans Autom Control* AC-20 218–226 and AC-22 328–337

[5] O'Malley R E 1991 *Singular Perturbation Methods for Ordinary Differential Equations* (Appl. Math. Sci. vol 89) (New York: Springer-Verlag) p 227

[6] Vasil'eva A B, Butuzov V F and Kalachev L V 1995 *The Boundary Function Method for Singular Perturbation Problems* (SIAM Studies in Appl. Math. vol 14) p 219 O'Malley R E, Jameson A 1975-77 Singular perturbations and singular arcs I, II. *Trans. Autom. Control.* **AC-20** 218–226 and **AC-22** 328–337

[7] Shchepakina E, Sobolev V and Mortell M P 2014 *Singular Perturbations. Introduction to System Order Reduction Methods with Applications* (Lect Notes in Math **2114**) (Berlin–Heidelber–London: Springer)

[8] Sobolev V A 2016 Slow Integral Manifolds and Control Problems in Critical and Twice Critical Cases *J. Phys.: Conf. Ser.* **727** 012017

[9] Strygin V V and Sobolev V A 1976 Effect of geometric and kinetic parameters and energy dissipation on orientation stability of satellites with double spin *Cosmic Research* **14** 331–335

[10] Kononenko L I and Sobolev V A 1994 Asymptotic expansion of slow integral manifolds *Siberian Math J* **35** 1119–1132

[11] Kokotović P V, Khalil K H and O'Reilly J 1986 *Singular Perturbation Methods in Control: Analysis and Design* (Philadelphia: SIAM)

[12] Smetannikova E and Sobolev V 2005 Regularization of Cheap Periodic Control Problems. *Autom. Remote Control* **66:6** 903–916

[13] Osintsev M S and Sobolev V A 2014 Reduction of dimension of optimal estimation problems for dynamical systems with singular perturbations *Comp Math. and Math. Phys.* **54** 45–58

[14] Osintsev M S and Sobolev V A 2013 Dimensionality reduction in optimal control and estimation problems for systems of solid bodies with low dissipation *Autom. Remote Control* **74:8** 1334–1347

[15] Sobolev V A 1991 Singular perturbations in linearly quadratic optimal control problems *Autom. Remote Control* **52:2** 180–189

[16] Panovko Y G 1967 *Fundamentals of applied theory of elastic vibrations* (Moscow: M.).

IOP Conf. Series: Journal of Physics: Conf. Series **811** (2017) 012010 doi:10.1088/1742-6596/811/1/012010

Order reduction for a model of marine bacteriophage evolution

Silvia Pagliarini[1] and Andrei Korobeinikov[2,3]

[1]Univestità degli Studi di Verona, Verona, Italy
[2]Centre de Recerca Matemàtica, Campus de Bellaterra 08193 Bellaterra, Barcelona, Spain
[3] Departament de Matemàtiques, Universitat Autònoma de Barcelona, Campus de Bellaterra, Edifici C, 08193 Bellaterra, Barcelona, Spain

E-mail: [1]`silvia.pagliarini92gmail.com`, [2]`akorobeinikov@crm.cat`

Abstract.
A typical mechanistic model of viral evolution necessary includes several time scales which can differ by orders of magnitude. Such a diversity of time scales makes analysis of these models difficult. Reducing the order of a model is highly desirable when handling such a model. A typical approach applied to such slow-fast (or singularly perturbed) systems is the time scales separation technique. Constructing the so-called quasi-steady-state approximation is the usual first step in applying the technique. While this technique is commonly applied, in some cases its straightforward application can lead to unsatisfactory results.

In this paper we construct the quasi-steady-state approximation for a model of evolution of marine bacteriophages based on the Beretta-Kuang model. We show that for this particular model the quasi-steady-state approximation is able to produce only qualitative but not quantitative fit.

1. Introduction

Mathematical modelling of biological evolution is a challenging task that modern mathematical science has to address. Due to their comparative simplicity, viruses can serve as an excellent model in evolutionary biology and the most convenient object for such modelling. However, constructing a mechanistic (that is, based on first principles) model and dealing with a typical mechanistic model in evolutionary biology can be a highly non-trivial task. The reason for this is that any mechanistic mathematical model of biological evolution, including models of viral evolution, must comprise a number of factors, including replication of all species involved, random mutations and a mechanism of natural selection. The time scales, on which these factors act, can vary in orders of magnitude. Thus, a mechanistic model of viral evolution can include the

IOP Conf. Series: Journal of Physics: Conf. Series **811** (2017) 012010 doi:10.1088/1742-6596/811/1/012010

dynamics of uninfected and infected target cells, as well as dynamics of virus. An average time scale of the uninfected target cells can be of order of days, whereas the life-span of free virus is of order of hours. These scales have to be compared with evolution, which typically is an extremely slow process. It is this diversity of time scales that makes analytical and numerical study of these systems difficult. That is, a typical mechanistic (that is, based on the first principles) model of biological evolution is unavoidably a slow-fast, or singularly perturbed, system.

Reduction of the order is highly desirable for this kind of models. A typical technique of order reduction is the so-called time scales separation technique. However, straightforward application of this technique, and in particularly constructing the so-called quasi-steady-state approximation, can lead to disappointing results.

In this paper we apply this technique to a model of marine bacteriophage evolution and construct a quasi-steady-state approximation of this model. Results of simulation show, that for this particular model the quasi-steady-state approximation does not provide a sufficiently accurate approximation, and terms of higher order are necessary.

2. Model of bacteriophage evolution

We consider the following model of bacteriophage evolution, which is introduced in [5]:

$$\begin{aligned}
\frac{dS(t)}{dt} &= \alpha S(t)\left(1-\frac{1}{C}\left(S(t)+\int_{\Omega} I(r,t)dr\right)\right)-\int_{\Omega} k(r)P(r,t)S(t)dr,\\
\frac{\partial I(r,t)}{\partial t} &= k(r)P(r,t)S(t)-\lambda(r)I(r,t)+q\Delta I(r,t),\\
\frac{\partial P(r,t)}{\partial t} &= -k(r)P(r,t)S(t)-\mu(r)P(r,t)+b(r)\lambda(r)I(r,t).
\end{aligned} \tag{1}$$

This model is based upon the Beretta-Kuang model of marine bacteriophage dynamics [1] and describes interaction of three populations, namely susceptible target bacteria, infected target bacteria and free bacteriophages. The Beretta-Kuang model postulates that the target bacteria reproduce according to the logistic law with reproduction rate α and carrying capacity of the environment C. The susceptible bacteria are infected by the phages in free infective stage at a bilinear rate with coefficient k. Immediately after an instance of infection, the infected bacterium moves into infected bacteria compartment. An average duration of the infective stage is $1/\lambda$, and then the infected bacterium dies. At the moment of death the bacterium released on average b free bacteriophages.

In contrast to the original Beretta-Kuang model where populations in all three compartments are postulated identical, model (1) assume the existence of multiple variants of the bacteriophage and postulate that these are continuously distributed in a continuous variant (phenotype) space Ω. Each of these variants is characterized by its phenotype. In the Beretta-Kuang model, a phage is characterized by four parameters, namely an average number of virions produced by an infected bacterium, b, an average life span of an infected bacterium, $1/\lambda$, an average free phage life span, $1/\mu$, and per capita rate of infection, k. Each of the viral variants in model (1) is characterized by a corresponding set of parameters k, λ, μ and b as well. In a model of phage evolution, any of these parameters can be variant-dependent, and, hence, the continuous variant space Ω is of dimension up to 4. (Please note that carrying capacity of the environment C and target bacteria per capita reproduction rate α do not depend on viral type, and, therefore, are constant.)

The model is comprises three variables: the susceptible bacterial population $S(t)$ (which is phenotype independent), density distribution in the variant space Ω of free phage population $P(r,t)$, and density distribution of infected bacteria $I(r,t)$ according to viral variants they are infected with. Random mutations in the continuous variant space are modelled by dispersion operator $q\Delta I \equiv q\frac{\partial^2 I}{\partial r^2}$, where coefficient q is assumed to be proportional to a probability of mutation, and is constant and small.

For the sake of simplicity, we assume that the only parameter depending on a phenotype is $k = k(r)$, whereas λ, μ and b are variant-independent (and hence are constant). This allows to reduce dimension of the variant space to 1-dim. For convenience, we assume the variant space Ω is the positive semi-axis of a 1-dimensional real space, $\Omega = \mathbb{R}^1_+ = (0, +\infty)$, and $r \in \Omega$ is the coordinate in this space.

3. Model order reduction

An intrinsic property of model (1), which motivates this study is that the model includes at least four different time scales, namely (i) life-span of the susceptible bacteria, (ii) life-span of the infected bacteria, (iii) life-span of the free phages, and, finally, (iv) a characteristic time scale of evolution. These time scales may differ by orders of magnitude: for instance, an average time for division of the susceptible cells is of order of a day, whereas the life-span of free virus is of order of hours. Such a diversity of time scales makes analytical and numerical study of this system difficult. For such a model, reducing its order is highly desirable. The reduction can be done using a time scales separation technique, which allows to reduce the considered system of three equations to one or two equations. In this section we construct a quasi-steady-state approximation of the system, applying the technique developed in [2, 4] for a similar model of RNA virus evolution and the original Beretta-Kuang model.

Let us denote $t = T\bar{t}$ and $r = R\bar{r}$, where T and R are constant scales that have to be defined, whereas $\bar{t}$ and $\bar{r}$ are non-dimensional variables. Furthermore, we denote $S(T\bar{t}) = \tilde{S}\bar{S}(\bar{t})$, $I(T\bar{t}, R\bar{r}) = \tilde{I}\bar{I}(\bar{t},\bar{r})$ and $P(T\bar{t}, R\bar{r}) = \tilde{P}\bar{P}(\bar{t},\bar{r})$, where $\tilde{S}$, $\tilde{I}$ and $\tilde{P}$ are also constant scales, while $\bar{S}(\bar{t})$, $\bar{I}(\bar{t},\bar{r})$ and $\bar{P}(\bar{t},\bar{r})$ are non-dimensional variables. Let us assume that

$$T = 1/q, \qquad R = 1, \qquad \tilde{S} = C, \qquad \tilde{P} = b\lambda\tilde{I}/\mu, \qquad \tilde{I} = C.$$

IOP Conf. Series: Journal of Physics: Conf. Series **811** (2017) 012010 doi:10.1088/1742-6596/811/1/012010

Substituting these variables into the system (1), we obtain non-dimensional system

$$
\begin{aligned}
\epsilon\frac{d\bar{S}(\bar{t})}{d\bar{t}} &= \bar{S}(\bar{t})\left(1-\bar{S}(\bar{t})-\int\limits_{\Omega}\bar{I}(\bar{t},\bar{r})d\bar{r}\right)-\frac{b\lambda C}{\mu\alpha}\bar{S}(\bar{t})\int\limits_{\Omega}\bar{P}(\bar{t},\bar{r})k(\bar{r})d\bar{r},\\
\frac{\partial\bar{I}(\bar{t},\bar{r})}{\partial\bar{t}} &= \frac{b\lambda C}{\mu q}k(\bar{r})\bar{P}(\bar{t},\bar{r})\bar{S}(\bar{t})-\frac{\lambda}{q}\bar{I}(\bar{t},\bar{r})+\Delta\bar{I}(\bar{t},\bar{r}),\\
\epsilon\nu\frac{\partial\bar{P}(\bar{t},\bar{r})}{\partial\bar{t}} &= -\frac{C}{\mu}k(\bar{r})\bar{P}(\bar{t},\bar{r})\bar{S}(\bar{t})-\bar{P}(\bar{t},\bar{r})+\bar{I}(\bar{t},\bar{r}).
\end{aligned}
\tag{2}
$$

Here $\epsilon = q/\alpha$ and $\nu = \alpha/\mu$ are small parameters. Indeed, ϵ is directly proportional to mutation rate q, which is assumed to be small, whereas ν is ratio of the characteristic time-scales of the free phages and the target bacteria. If ϵ and ν are small, the scale separation technique can be applied to this system.

To obtain a quasi-steady-state approximation of this system, we put $\nu = 0$ (see [2,4]). This yields the following reduced model:

$$
\begin{aligned}
\epsilon\frac{d\bar{S}(\bar{t})}{d\bar{t}} &= \bar{S}(\bar{t})\left(1-\bar{S}(\bar{t})\right)-\int\limits_{\Omega}\left(\bar{S}(\bar{t})+\frac{b\lambda}{\alpha}\frac{Ck(\bar{r})\bar{S}(\bar{t})}{Ck(\bar{r})\bar{S}(\bar{t})+\mu}\right)\bar{I}(\bar{t},\bar{r})d\bar{r},\\
\frac{\partial\bar{I}(\bar{t},\bar{r})}{\partial\bar{t}} &= \frac{\lambda}{q}\left(b\frac{Ck(\bar{r})\bar{S}(\bar{t})}{Ck(\bar{r})\bar{S}(\bar{t})+\mu}-1\right)\bar{I}(\bar{t},\bar{r})+\Delta\bar{I}(\bar{t},\bar{r}).
\end{aligned}
\tag{3}
$$

Here, for simplicity, it is assumed that only one parameter, namely $k = k(r)$, depends on a viral variant, whereas the others parameter are assumed to be constant. Furthermore, here $\bar{r} \in (0, +\infty)$.

4. Results

To explore how well the quasi-steady-state approximation (3) approximates the original model (1), we run simulations for both these models. In these simulations we use the following values for the system parameters: $q = 10^{-6}$ day^{-1}, $\alpha = 1.5$ day^{-1}, $C = 100$ cells, $\lambda = 0.3$ day^{-1}, $\mu = 2$ day^{-1} and $b = 14$. For the sake of simplicity, we assume that $k(r)$ is linearly depending on r, $k(r) = \xi r$, where $\xi = 0.002$ day^{-1}. The value of ξ is chosen in such a way to ensure that the basic reproduction number $R_0 = (b-1)k(r)C/\mu$ is equal to 1 at some r. In simulations we assume that $r \in (0, r_{end})$ and set no-flux boundary conditions for $P(r,t)$ and $I(r,t)$ at $r = 0$ and $r = r_{end}$. For model (1), we set initial conditions $S(0) = C$ and $P(r,0) = 0$, and assume that $I(r,0)$ differs from zero only in a narrow vicinity of $r = 1$. For model (3) we use the corresponding re-scaled initial conditions $\bar{S}(0) = 1$ and $\bar{I}(\bar{r},\bar{0}) = I(r,0)/C$.

Figure 1 shows distributions of the infected bacteria by viral variants which they are infected with in the viral variant space Ω for for model (1) (the left-hand panel) and its quasi-steady-state approximation (3) (the right-hand panel, respectively). (In Fig. 1, colours corresponds to the density of infected bacteria $I(r,t)$; see legends at the right-hand side of each panel). The formation of a pulse-type travelling wave moving in the right-hand direction in the variant space, that is towards higher Darwinian fitness, is clearly seen for both, the original system and its quasi-steady-state approximation. Please also note that the speed of evolution (the speed of the traveling waves) in Fig. 1 is not constant for either of the systems: it can be seen that evolution goes faster as the Darwinian fitness grows, and then, when some level of the fitness is reached, it abruptly slows down and remans approximately constant thereafter. It is noteworthy that the same effect was observed in [3].

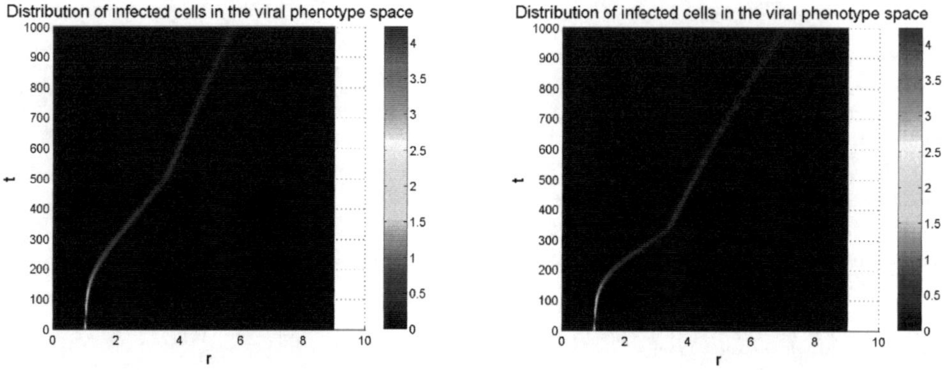

Figure 1. Distribution of the infected bacteria by viral type they are infected with in the variant space in time for models (1) (left-hand panel) and (3) (right-hand panel). Here $q = 10^{-6}$, $\alpha = 2$, $C = 100$, $\lambda = 3$ $\mu = 20$, $b = 149.254$, and $\xi = 0.002$. The colours correspond to the density; see the legend on the right-hand sides of both panels. Please note the formation of a pulse-type traveling wave moving in the right-hand direction in the variant space. Please also note that the speed of the waves in both panels is not constant and different for both models.

There is a reasonably good qualitative fit between the data for both models in the sense that the qualitative behaviour of both models is nearly the same. However, the simulations also reviles that the evolution goes faster in the quasi-steady-state approximation than it does in the original model. The disparity in the speed of evolution between the original model and its quasi-steady-state approximation is also seen in Figures 2 and 3. Figure 2 shows distribution of the infected cells by viral variants they are infected with for model (1) (curve #1) and its quasi-steady-state approximation (3) (curve #2) at time $t = 1000$. It can be seen that that the median of distribution (2) is located notably further to the right-hand side than that of distribution (1). This disparity can be explained only by a faster evolution exhibited by the quasi-steady-state approximation (3) compared with this of the original model (1).

Figure 3 shows the dynamics of the total infected population $\int_{\Omega} I(r,t)dr$ (left-hand panel) and susceptible population $S(t)$ (right-hand panel) in time. In both these pictures, curves #1 correspond to model (1) and curve #2 corresponds to its quasi-steady-state approximation (3). These pictures also show that, while the qualitative behaviour of both models is very similar, the evolution in the quasi-steady-state approximation progresses faster.

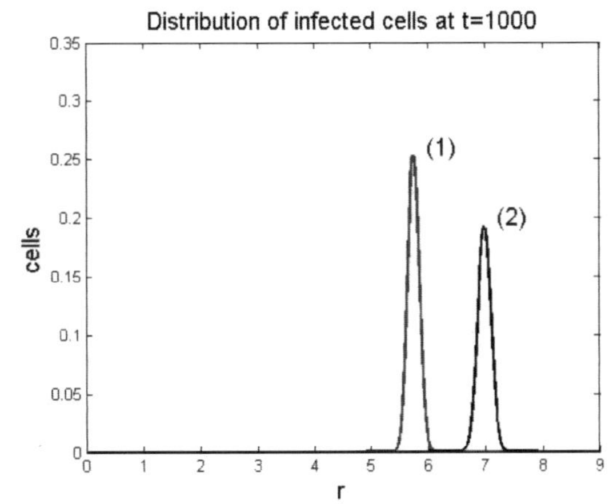

Figure 2. Distributions of infected bacteria by the viral variants they are infected with $I(r,t)$ for model (1) (curve #1) and its quasi-steady-state approximation (3) (curve #2) at time $t = 1000$. Please note that the distribution (2) is located further to the right-hand side than distribution (1). This difference reflects the higher speed of evolution that is exhibited by the quasi-steady-state approximation.

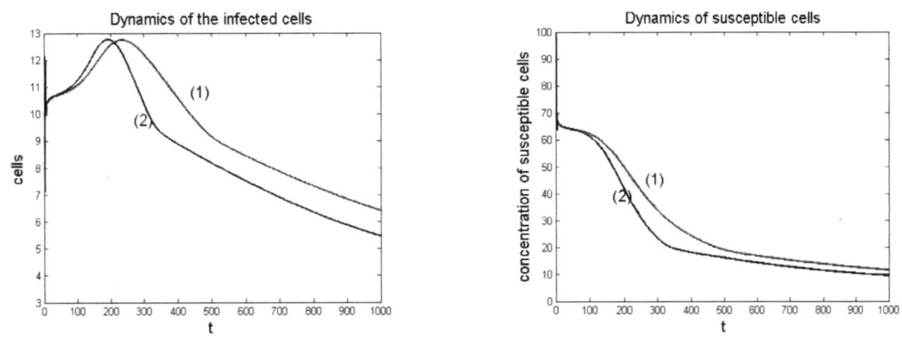

Figure 3. Dynamics of the infected ($\int_\Omega I(r,t)dr$) and susceptible bacteria for model (1) (curves #1) and its quasi-steady-state approximation (3) (curves #2) in time. Here $q = 10^{-6}$, $\alpha = 2$, $C = 100$, $\lambda = 3$ $\mu = 20$, $b = 149.254$ and $\xi = 0.002$.

An intrinsic property of the Beretta-Kuang model is that, despite the bilinearity of the infection rate, it exhibits a admits a supercritical Hopf bifurcation, with formation of a stable limit cycle in the model phase space [1,4]. The dynamics of the susceptible population for model (1) and its quasi-steady-state approximation for parameters, which corresponds to the presence of the stable limit cycle in the original model, is shown in Figure 4. Please note the different behaviour of the evolution model and its quasi-steady-state approximation: it is easy to see in these pictures that in both panel curves (2), that correspond to data for the quasi-steady-state approximation, tend to an equilibrium values, whereas curves (1), which correspond to the original model, revile the presence of a stable limit cycle in the system. That is, the Hopf bifurcation occurs at different parameter values in the original model and in its quasi-steady-state approximation. The same observation was reported in [4].

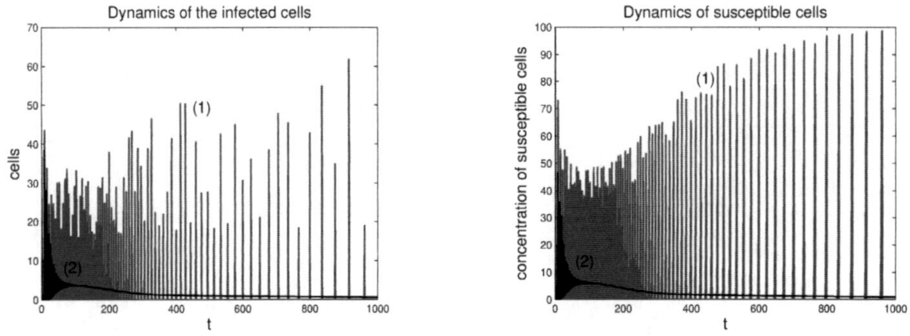

Figure 4. Dynamics of the infected ($\int_\Omega I(r,t)dr$) and susceptible bacteria for model (1) (curves #1) and its quasi-steady-state approximation (3) (curves #2) in time. Here $q = 10^{-6}$, $\alpha = 2$, $C = 100$, $\lambda = 3$ $\mu = 20$, $b = 149.254$, and $\xi = 0.02$.

A general conclusion that can be withdrawn from these results is that for this particular model the quasi-steady-state approximation does not provide satisfactory quantitative results matching to these for the original model, and hence terms of higher orders should be taken in account in an reduced model.

5. Conclusions

In this paper we attempted to make a model order reduction for a model of evolution of aquatic bacteriophage, which is based on the Beretta-Kuang model of bacteriophage dynamics. For this we constructed a quasi-steady-state approximation of this model. In many application, constructing such an approximation is sufficient for model order reduction. However, for this particular model, comparison of results for the original model and its quasi-steady-state approximation reviles that the approximation does not produces sufficiently good match to the original model: while qualitative behaviour of both, the model and its approximation, matches, the quantitative outcomes significantly differ. Thus, the approximation produces a notably faster evolution that the original model does. Moreover, Hopf bifurcation appears for different parameter value. The latter observation coincides with observations in [4], where it was shown that a quasi-steady-state approximation of the Beretta-Kuang ODE model, which is the basis for the model of viral evolution considered in this paper, does not preserve totally properties of the Beretta-Kuang model. In general, these results demonstrate that a more accurate approximation, taking into consideration terms of higher order, can be necessary for constructing a reduced model of this particular model. Moreover, these results also show that the widely accepted approach, when the order reduction for a model is limited to construction of a quasi-steady-state approximation, should be applied with significant amount of care and reservations.

Acknowledgment

Andrei Korobeinikov is supported by the CERCA Programme of the Generalitat de Catalunya, by the Ministry of Economics of Spain via Ramón y Cajal Fellowship RYC-2011-08061 and grant MTM2015-71509-C2-1-R, by AGAUR of the Generalitat de Catalunya via grant 2014SGR-1307 and by CONACYT (México) via grant 219614.

References

[1] Beretta E and Kuang Y 1998 Modeling and analysis of a marine bacteriophage infection. *Math. Biosci.* **149** 57–76.

[2] Korobeinikov A, Archibasov A, Sobolev V 2015 Order reduction for a RNA virus evolution model, *Math. Biosci. Eng.* **12** 1007–1016.

[3] Korobeinikov A and Dempsey C 2014 A continuous phenotype space model od RNA virus evolution within a host. *Math. Biosci. Eng.* **11** 919–927.

[4] Korobeinikov A, Shchepakina E and Sobolev V 2015 Paradox of enrichment and system order reduction: bacteriophages dynamics as case study. *Math. Med. Biol.* (Advance Access pub.) doi:10.1093/imammb/dqv025

[5] Pagliarini S and Korobeinikov A 2016 A mathematical model of marine bacteriophage evolution Math. Biosci. Eng. (submitted) CRM Preprint

IOP Conf. Series: Journal of Physics: Conf. Series **811** (2017) 012011 doi:10.1088/1742-6596/811/1/012011

A model of hysteresis arising from social interaction within a firm

Luis A Rios
Wharton School of Business, University of Pennsylvania, USA
E-mail: luisrios@wharton.upenn.edu

Dmitrii Rachinskii
Department of Mathematical Sciences, The University of Texas at Dallas, USA
E-mail: dmitry.rachinskiy@utdallas.edu

Rod Cross
Department of Economics (*Emeritus*), University of Strathclyde, UK
E-mail: rod.cross@strath.ac.uk

Abstract. This paper subscribes to the view that a key distinguishing feature of a firm is its social nature. We present a model in which hysteresis arises from the social interactions between employees. Employees have a simple response to incentives in the form of the pay available outside the firm relative to that available within the firm. Allowing for social interaction, whereby employees are influenced by the effort levels of fellow employees, leads to the distinctive effects, such as lazy relay responses to incentives, associated with hysteresis.

1. Introduction

There is a huge literature on why some economic activities are conducted by way of "markets", and some by "firms". In this paper we follow the account of "the" firm that stresses the social nature of the interactions between the employees in that mode of organizing economic activity. Social relationships obviously pertain to markets as well. Think of the social interactions between buyers and sellers in an auction market, for example, as presided over by the entity conducting the auction, which is itself often a firm. There is a spectrum here, with the firm involving closer, more continuous, social contacts between employees than is the case with individuals interacting during market transactions: the "no person is an island" precept applies more strongly to firms than to markets.

The strongly social nature of the interactions between employees has been taken to be the distinctive feature of firms in a variety of accounts. Simon has stressed the willingness to accept authority by a firms employees [1]; Bowles the ability of a firm to shape the private preferences of its employees into social ones [2]; Casadesus-Masanell the ability of a firm to imbue a work ethic in employees [3]; Holmstrom and Milgrom the incentives to multitasking by employees within a firm [4]; and finally, but by no means exhaustively, Akerlof and Kranton have stressed the sense of identity, or belonging, which a firm can engender in its employees [5].

In the present paper we take as given that such social forces are central to the fabric of a firm [6] and ask what happens as a result of such social interactions within a firm. Our conjecture is that social interactions within a firm give rise to hysteresis in the way the employees within the firm respond to the incentives offered within the firm relative to those available outside the firm. A manifestation of this would be that there is a range of relative pay incentives within which employees will not change their behavioral responses, in the form of quitting the firm, for example.

The term hysteresis comes from the Greek to be late or come behind, and was coined by Ewing to describe the behavior of electromagnetic fields in ferric metals [7]. If a magnetizing force is applied to such metals, then reversed, the metals do not revert to their previous characteristics: they remain changed. As a general systems property [8], hysteresis implies that the outputs of a system do not just depend on the contemporaneous values of the inputs applied, but also retain a memory, such as of the non-dominated extremum values of the inputs applied in the past. Thus there are effects that remain after the immediate causes are removed: history is not bunk. If our conjecture that the social interactions within a firm lead to hysteresis is correct, the firm will display a form of temporary inertia, in that employees participation decisions will not respond to all changes to relative pay incentives; and the pool of employees who opt to remain working within the firm will be shaped by a memory of the relative pay incentives experienced in the past.

The rest of this paper is structured as follows. In the next section we outline a model of hysteresis, which arises when employees, who have the same responses to relative pay rate incentives, can observe the effort levels of fellow employees. We then construct probabilistic versions of this model and discuss their dynamic properties.

2. A model of hysteresis within a firm

Initially, the term hysteresis was coined to describe results obtained in experiments on electromagnetic fields in ferric metals. The need for this new term was felt by Ewing because he thought that hysteresis effects would apply to a wider range of phenomena [7]. This prophecy has been borne out in the wide range of areas of inquiry in which hysteresis has been discovered, ranging from physics to biology, materials science to mechanics and from electronics to economics (see the entries in [9]). In the literature on firms Ford speculated that organizations might display hysteresis in managerial intensity when they move from expansionary to contractionary modes [10], but there appears to have been little by way of follow up to this conjecture in the literature on organizations or management. Ford suggested that participation decisions were one possible reason why organizational hysteresis might arise, but did not offer a formal model of such decisions.

The present paper aims to fill this gap by providing a formal model of how hysteresis can arise in a firm as a result of the social interactions involved when employees can observe the effort levels of their fellow employees.

The formalization of hysteresis as a general systems property was made by Krasnosel'skii and Pokrovskii [8]. The most widely used model of hysteresis is that derived from the work of Preisach [11], in which hysteresis at the aggregate or systems level arises from elements at the micro level that are heterogeneous and respond non-linearly to a common input shock. In such a world the aggregate output of the system displays remanence, in that the application and removal of an input shock will not see the systems output return to the status quo ante; and the output of the system displays an erasable, selective memory of the non-dominated extremum values of the shocks experienced.

A widely used specification of non-linear responses at the micro level is the non-ideal or lazy relay. In such a relay the individual elements do not respond to an input shock until some value, α, is reached, in which case the behavior, changes from some state to another, say low to high; and does not switch from the high to the low state until some value of the input shock, β, is reached, $\beta < \alpha$. Thus there is a range of values for the input i, $\beta < i < \alpha$, inside which behavior will not change in the face of variation in i. Hence in order to know whether the individual element is in a high or low state it is necessary to know the history of i, which explains whether the range $\beta < i < \alpha$ has been approached from above or below. Heterogeneity can be introduced by allowing the α and β switching points to be different between the elements. The systems output y would then depend not just on the contemporaneous value of i, but also on the non-dominated extremum values of i experienced in the past [12].

The model of hysteresis proposed in the present paper has a more parsimonious specification of the elements, in this context employees, in the system. We assume that employees are homogeneous in the sense that they respond in the same way to an incentive in the form of the pay rate offered by the firm relative to what they could earn by not being employed by the firm. The rate of pay that pertains outside the firm is held constant, the relative pay incentive being varied by changing the rate of pay offered to employees within the firm. The other parsimonious feature is to employ a Heaviside step function, instead of a non-ideal or lazy relay, to describe how the employees initially respond to relative pay incentives. Thus there is a single trigger α, for relative pay, with values of relative pay greater than α, leading the individuals to choose to be employed within the firm; and values of relative pay less than or equal to α, leading the individuals to choose not to be employed by the firm. The intriguing result is that if employees within the firm can observe the effort levels of other employees, and react by raising/lowering their effort levels in sympathy with the effort levels of fellow employees, hysteresis arises. This hysteresis takes the form of a lazy relay between incentives and the number of individuals who elect to remain employed within the firm. If this lazy relay applies, the way would then be open to use Preisach-type models to analyze conglomerate firms as composed of operational units characterized by such processes.

3. Hysteresis emerging from interaction

The following construction can be seen as a prototype model for emerging hysteretic response that results from interaction of non-hysteretic responses [13]. More complex systems of interacting agents will produce more complex hysteric response to exogenous stimuli in a similar fashion [14].

Consider two identical agents which respond to incentive i according to the simple threshold rule. Each agent, at any given time, adopts one of the two states: either the high performance state $s = 1$ or the low performance state $s = 0$. When $i \leq \alpha$, the agents adopt the state $s = 0$; when $i > \alpha$, they adopt the state $s = 1$. In other words, if we introduce the Heaviside step function

$$H_\alpha(x) = \begin{cases} 0, & x \leq \alpha \\ 1, & x > \alpha \end{cases},$$

we simply have

$$s = H_\alpha(i) \tag{1}$$

at all times, where s is the state of either agent. Here α is a threshold, which presents a sufficient level of incentive for an agent to adopt the high performance state. As i (the input) is varied, both agents switch from one state to the other each time the input crosses the threshold value α, see Fig. 1(b). There is no hysteresis or memory in this response. A counterpart of such response, which allows the performance to increase continuously with incentive, can be modeled by the piecewise linear function

$$s = \begin{cases} 0, & i \leq \alpha \\ k^{-1}(i-\alpha), & \alpha < i < \alpha + k \\ 1, & i \geq \alpha + k \end{cases}, \tag{2}$$

see Fig. 1(a).

The above agents act independently. However, if agents affect each other, hysteresis may emerge. Suppose that the "percepted incentive" (or, the total incentive) for the first agent is

$$\tilde{i}_1 = i + ks_2,$$

where $k > 0$ is some coefficient, which is smaller than α. Similarly, assume that the total incentive for the second agent is

$$\tilde{i}_2 = i + ks_1.$$

Now, the switching rules become

$$s_1 = H_\alpha(\tilde{i}_1), \qquad s_2 = H_\alpha(\tilde{i}_2).$$

In other words, the response of the agents to variations of the applied incentive i is determined by the implicit coupled equations

$$s_1 = H_\alpha(i + ks_2), \qquad s_2 = H_\alpha(i + ks_1). \tag{3}$$

We need to find a solution s_1, s_2 of this system as a "function" of the given input i (although, as a matter of fact, this is not a function, but an operator with memory).

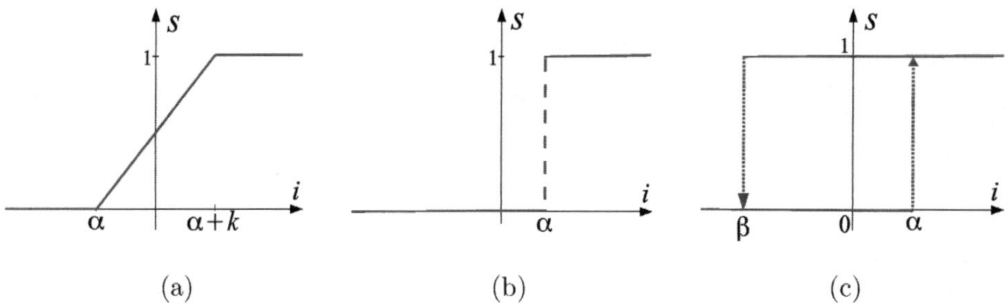

Figure 1. (a) Piecewise linear response. (b) Heaviside step function. (c) Non-ideal relay response (rectangular hysteresis loop).

The first observation is that if $s_1 = s_2$ at some moment in time, then the states of the two agent will remain equal at all later times. Therefore, we can consider the common state $s = s_1 = s_2$ of the two agents, and the system reduces to one implicit equation

$$s = H_\alpha(i + ks). \tag{4}$$

It turns out that this equation defines the non-ideal relay with two thresholds, also known as an elementary rectangular hysteresis loop [13, 15]. That is, i is the input and s is the output of the non-ideal relay, with s switching from 0 to 1 and backwards at two different threshold values of i, see Fig. 1(c).

Indeed, assume that initially $i = 0$ and $s = 0$, and let i increase. As i exceeds the value α, the right hand side becomes equal to 1, hence the state s becomes $s = 1$. That is, one switching threshold remains equal to α. From this moment (and until the state switches to zero again), the input $\tilde{i} = i + ks$ of H_α becomes equal to $\tilde{i} = i + ks = i + k$, hence $s = H_\alpha(i + k)$. The right hand side will become zero when the sum $i + k$ reaches the value α, i.e. when i reaches the value $\alpha - k$. Hence, the second switching threshold is $\beta = \alpha - k$. There is a difference between the switching thresholds for the transitions from 0 to 1 and backwards, and this difference equals to k. As k measures the strength of the affect of one agent on the other, we can say that a stronger interaction results in a larger separation of thresholds in the response—the more interaction, the more hysteresis.

4. A stochastic switching model

It seems realistic to assume that thresholds in individuals are not deterministic, but rather transitions of an individual between the states are a stochastic process. That is, given a certain input (e.g. pay rate), switching the state is a random event characterized by a certain transition probability rate. Here we develop the model described in the previous subsection to include uncertainty in switching events. A simple Markov chain model with transition probability rates depending on the pay rate i is suitable for our purposes.

Assume that the probability rate for the transition from state $s = 0$ to state $s = 1$ for an individual is $p_{0,1}(i)$. This probability rate should naturally be an increasing function of i. Similarly, let us assume that the probability rate $p_{1,0}(i)$ for the transition from state $s = 1$ to $s = 0$ decreases with i. We treat i as a control variable (input) that can change with time, i.e. $i = i_t$, hence the transition probability rates become functions of time. This defines a continuous time Markov chain with two states and time-dependent transition probability rates. Denoting by π_t^s the probability to find the agent in state s at time t, we can write the master equation for the evolution of the probability distribution vector $\pi_t = (\pi_t^0, \pi_t^1)$:

$$\frac{d\pi_t}{dt} = \pi_t P(i_t), \tag{5}$$

where

$$P(i) = \begin{pmatrix} -p_{0,1}(i) & p_{0,1}(i) \\ p_{1,0}(i) & -p_{1,0}(i) \end{pmatrix} \tag{6}$$

and $\pi_t^0 + \pi_t^1 = 1$ at all times.

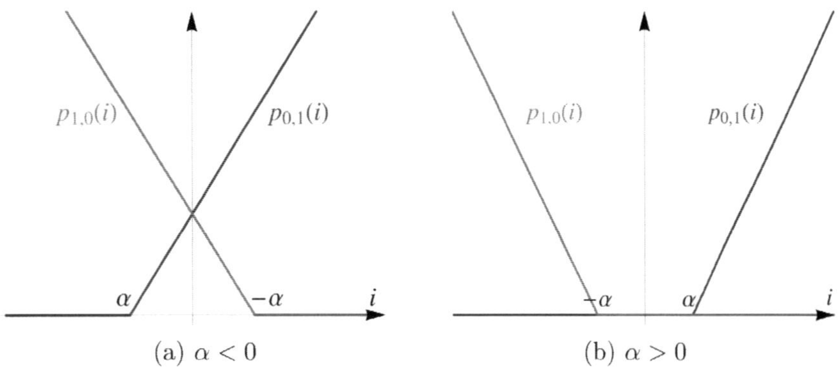

Figure 2. Probability of transition from state $s = 0$ to state $s = 1$ (blue) and vice versa (red) as a function of the stimulus i for one individual.

Next, we need to define the functions $p_{0,1}(i)$ and $p_{1,0}(i)$. As before, we assume that there are no transitions from state $s = 0$ to state $s = 1$ if the pay rate stays below a threshold value α. For the values of i exceeding α, we simply assume that the transition probability increases linearly with i. This leads to the law

$$p_{0,1}(i) = \gamma\,(i - \alpha) H_\alpha(i) \tag{7}$$

with the steepness coefficient (slope) $\gamma > 0$. Again, for simplicity, we assume symmetry in the law of transitions between the states by setting

$$p_{1,0}(i) = p_{0,1}(-i) = \gamma\,(-i - \alpha) H_\alpha(-i), \tag{8}$$

see Fig. 2. The stochastic model (5)–(8) with $\alpha \le 0$ replaces and extends the simple deterministic switching rule (1). The non-positive α corresponds to the absence of hysteresis in individuals. The same model with $\alpha > 0$ extends the deterministic switching rule based on the non-ideal relay [8]. That is, a positive α corresponds to the hypothesis that hysteresis is present in the response of individuals to the incentive, for example, due to interaction of individuals within a firm.

The model (3) of two interacting agents can be extended to include uncertainty in decision making along the same lines. Here, we have a Markov chain with four states, $(0,0),(0,1),(1,0),(1,1)$, with the first digit denoting the state of the first agent and the second digit standing for the state of the second agent. Assuming that decision making of an agent is affected by the other agent, we define the transition probability rates from state $(0,s)$ to state $(1,s)$ and from state $(s,0)$ to state $(s,1)$ by

$$p_{(0,s),(1,s)}(i) = p_{(s,0),(s,1)}(i) = \gamma\,(i+ks-\alpha)H_\alpha(i+ks), \tag{9}$$

where s is either 0 or 1 (cf. (3)). Similarly, the probability rates for transitions from state $(1,s)$ to $(0,s)$ and from state $(s,1)$ to $(s,0)$ are defined by

$$p_{(1,s),(0,s)}(i) = p_{(s,1),(s,0)}(i) = \gamma\,(-i-ks-\alpha)H_\alpha(-i-ks). \tag{10}$$

Finally, we set

$$p_{(0,0),(1,1)}(i) = p_{(0,1),(1,0)}(i) = p_{(1,0),(0,1)}(i) = p_{(1,1),(0,0)}(i) = 0. \tag{11}$$

Then, the evolution of the four-dimensional probability distribution vector $\pi_t = (\pi_t^{(0,0)},\pi_t^{(0,1)},\pi_t^{(1,0)},\pi_t^{(1,1)})$ is determined by the master equation

$$\frac{d\pi_t}{dt} = \pi_t \hat{P}(i_t), \tag{12}$$

where the non-diagonal entries $p_{(s_1,s_2),(\sigma_1,\sigma_2)}$ of the 4×4 transition rate matrix $\hat{P}(i)$ are defined by equations (9)–(11) and the diagonal entries are determined by the condition that the row sum for each row of $\hat{P}$ is zero. System (12) serves as a probabilistic analog of the deterministic model (3).

5. Dynamics of the stochastic model

The following results illustrate the response of the stochastic model presented in the previous section to a periodically varying incentive and highlight the dependence of this response on the model and input parameters.

Fig. 3 presents time varying probabilities π_t^0 and π_t^1 to find an agent in state $s=0$ (black) and state $s=1$ (red), respectively, generated by master equation (5) with the periodic input $i=\sin t$ for a positive threshold value α and a steep slope γ.

Fig. 4 shows the response of model (5) to a slowly varying periodic input depending on the threshold parameter α (see Fig. 2). An agent spends part of the period in the "pure" state $s=0$ ($\pi^0=1$), part of the period in the "pure" state $s=1$ ($\pi^0=0$), and part in a "mixed" state. Each loop with $\alpha>0$ departs from the line $\pi^0=0$ at the point $i=\alpha$ and lands at the line $\pi^0=1$ for a slightly larger value of i; similarly, it departs from the line $\pi^0=1$ when $i=-\alpha$ and lands at the line $\pi^0=0$ for a slightly lower value of α. Hence, for $\alpha>0$, we observe an almost rectangular hysteresis loop (brown, red),

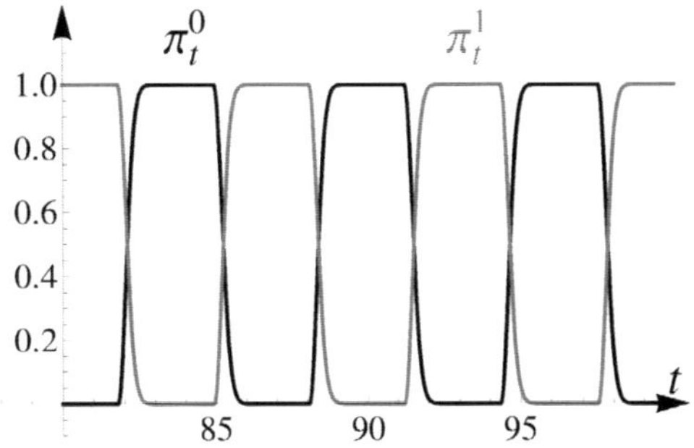

Figure 3. Periodic solution (π_t^0, π_t^1) of master equation (5) with the periodic input $i = \sin t$. Parameters are $\alpha = 0.1$, $\gamma = 15$.

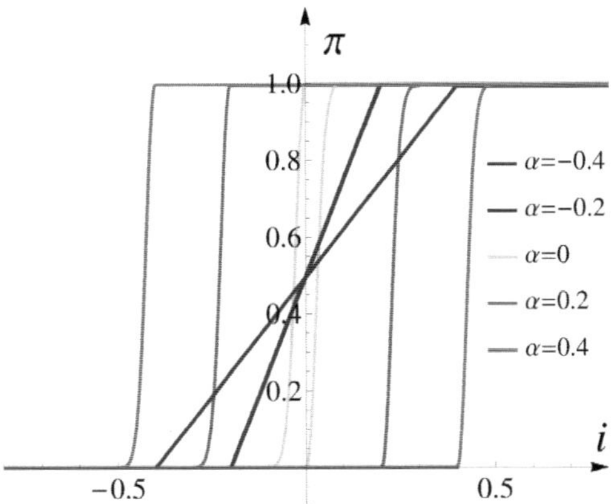

Figure 4. Input-output loops of model (5) (with $\gamma = 15$) generated by the slowly varying periodic input $i = \sin(0.01t)$ for different values of the threshold α.

which widens with α. For $\alpha < 0$, the loop degenerates to a piecewise linear curve (blue, purple) shown in Fig. 7(a), and no hysteresis is observed.

The shape of the input-output loop depends on the rate of change of the input, see Fig. 5 where the input is $i = 1.3\sin(wt)$. The left and right panels correspond to $\alpha < 0$ and $\alpha > 0$, respectively. One can see that the slower the input (smaller frequency w), the narrower the loop is. In particular, one can notice the "fake" hysteresis for higher frequencies on the left panel, which disappears for lower frequencies when the loop shrinks to almost a line (red). The "true" hysteresis is manifested by the low frequency loop (red) on the right panel. In the limit of low frequencies, the loop approaches the shape shown in Fig. 7—the piecewise linear (non-hysteretic) saturation function for $\alpha < 0$, and the rectangular hysteresis loop for $\alpha > 0$.

The slope γ of the transition probability rates (7) and (8) controls the shape of the

IOP Conf. Series: Journal of Physics: Conf. Series **811** (2017) 012011 doi:10.1088/1742-6596/811/1/012011

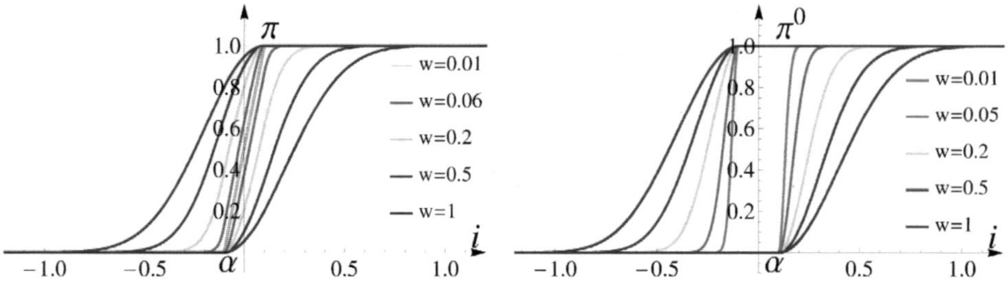

(a) $\alpha = -0.1$ (b) $\alpha = 0.1$

Figure 5. Input-output loops of model (5) (with $\gamma = 15$) generated by the periodic input $i = 1.3\sin(wt)$ for different frequencies w.

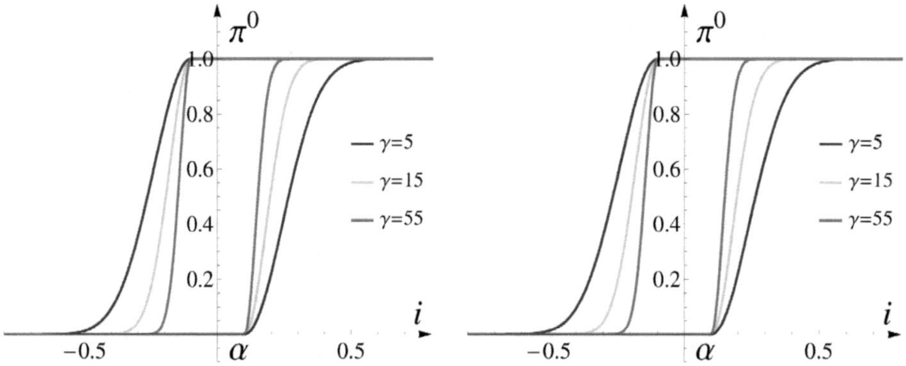

(a) $\alpha = -0.2$ (b) $\alpha = 0.2$

Figure 6. Input-output loops of model (5) generated by a periodic input $i(t) = \sin(0.1t)$ for different slopes γ of the functions $p_{01}(i), p_{10}(i)$ shown in Fig. 2.

loop in a similar fashion. As Fig. 6 shows, the steeper the slope γ, the closer the loop to the limit case shown in Fig. 7 is. That is, increasing γ results in the same effect as decreasing the input frequency w.

The limit shapes of the plots in Figs. 5 and 6 as $w \to 0$ and $\gamma \to 0$ are shown in Fig. 7 (cf. Fig. 1). For $\alpha < 0$, the probability π^0 to find an agent in state $s = 0$ is a piecewise linear function $\pi^0(i)$ shown on panel (a). There is no hysteresis. For $\alpha > 0$, the dependence of π^0 on i is described by the non-ideal relay operator (an elementary rectangular hysteresis loop) with two different thresholds $\pm\alpha$ for switching from state $s = 0$ ($\pi^0 = 1$) to $s = 1$ ($\pi^0 = 0$) and vice versa.

Fig. 8 presents the periodic solution of model (5) with the periodic input $i(t) =$

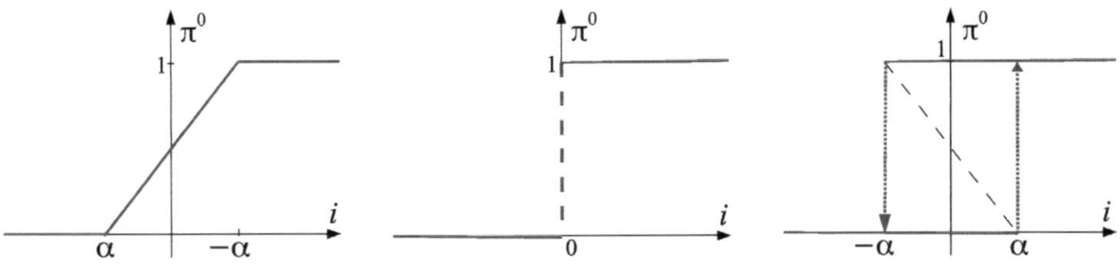

(a) $\alpha < 0$ (b) $\alpha = 0$ (c) $\alpha > 0$

Figure 7. Response of model (5) to a slowly varying input $i(t)$ in the limit when $\gamma \to \infty$ and $i'(t) \to 0$.

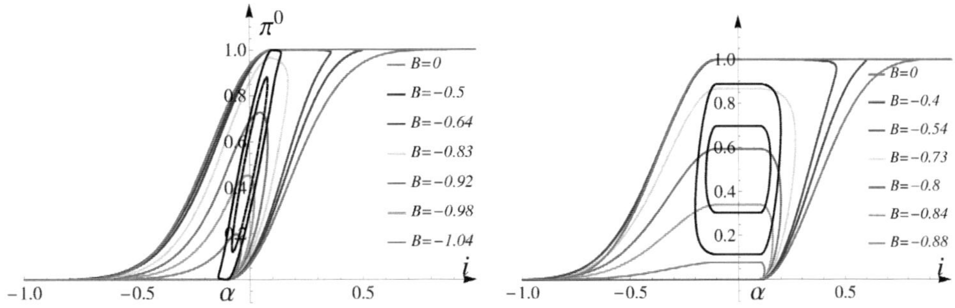

(a) $\alpha = -0.1$ (b) $\alpha = 0.1$

Figure 8. Input-output loops of model (5) (with $\gamma = 15$) generated by the periodic input $i(t) = B + R\sin t$ for different values of B.

$B + R\sin t$ for different values of B. The maximum and the minimum of the input are equal to $x_M = B + R$ and $x_m = B - R$, respectively. When $x_M < \alpha$, the agent resides at the pure state $s = 1$ ($\pi^0 = 0$) at all times. For $x_M > \alpha$, a loop appears and grows with x_M (which grows with B), see red, orange, brown, green curves on both panels. For larger values of B, the loop resides in each of the pure states $s = 0, 1$ for part of the period (blue, purple, and gray curves). Black loops that remain in the mixed state for all times, that is they do not touch the lines $s = 0, 1$, correspond to inputs with $B = 0$ and a sufficiently small amplitude R.

Finally, in Fig. 9, we consider the behavior of model (12) with two interacting agents for different values of the interaction strength k. The figure shows the dependence of the probability $\pi_t^{(1,1)}$ on the slowly varying input $i(t) = \sin(wt)$. The probabilities $\pi_t^{(0,1)}$ and $\pi_t^{(1,0)}$ are close to zero at all times, hence $\pi_t^{(0,0)} + \pi_t^{(1,1)} \approx 1$ for all t. In the case $\alpha < 0$

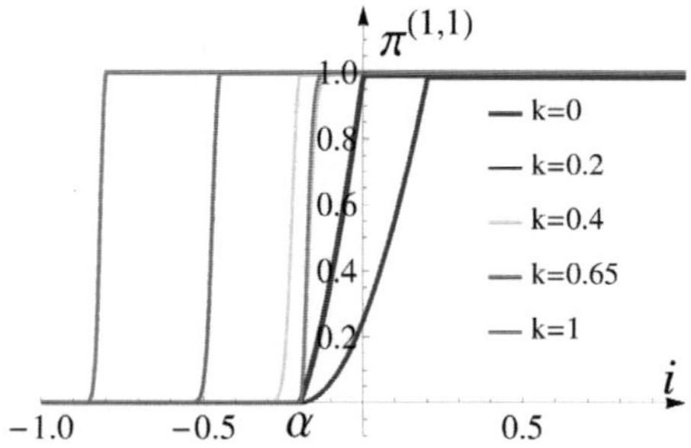

(a) $\alpha = -0.2$

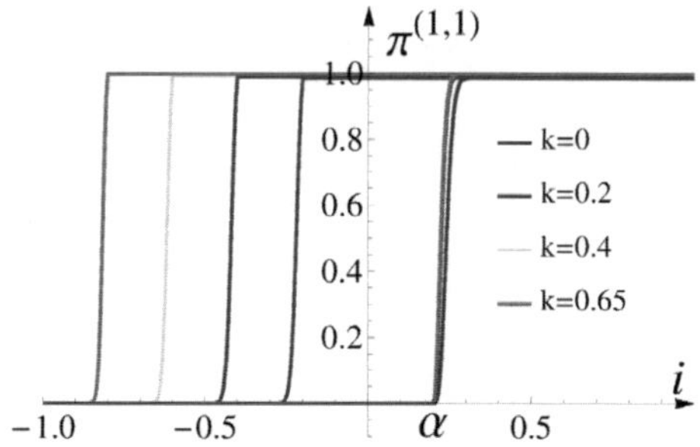

(b) $\alpha = 0.2$

Figure 9. Input-output loops of model (12) of two interacting agents generated for different values of the interaction strength k. The figure shows the dependence of the probability $\pi_t^{(1,1)}$ on the slowly varying input $i(t) = \sin(wt)$ with $w = 0.01$.

(left panel), for smaller values of k, the dependence is described by a sigmoid saturated function $\pi^{(1,1)}(i)$, no hysteresis is observed (blue and purple lines). For larger k, the hysteresis loop appears due to interaction of the agents. The right (ascending) segments of all the loops almost coincide, while the width of the loop grows with k (green, brown, and red lines). In the case $\alpha > 0$ (right panel), the hysteresis loop is present for all $k \geq 0$. Again the right segment of the loop is almost the same for all k, and the loop grows wider with increasing interaction strength k.

6. Conclusions

We have proposed a probabilistic switching system, which can potentially be used for modeling the behavioral response of employees to the incentives offered within a firm. Variants of the model include or ignore social interactions between the employees as a factor that can affect their behavior. In the limit of slow variations of the input, the model either reduces to a memoryless instantaneous functional relationship between the input and the output or gives rise to hysteresis in the form of a non-ideal relay operator (rectangular hysteresis loop), depending on the parameters used. In particular, a sufficiently strong interaction between the agents can induce hysteresis, which is not observed in non-interacting individuals. On the other hand, if hysteresis is present in individuals, the interaction can enhance it. Faster variations of the input produce wider input-output loops, which look like a hysteretic response but may shrink to a memoryless functional response curve as the rate of input variations slows down.

Acknowledgments

DR acknowledges the support of NSF through grant DMS-1413223.

References

[1] Simon H A 1991 Organizations and markets *Journal of Economic Perspectives* **5** 25–44
[2] Bowles S 1998 Endogenous preferences: The cultural consequences of markets and other economic institutions *Journal of Economic Literature* **36** 75–111
[3] Casadesus-Masanell R 2004 Trust in agency *Journal of Economic Management Strategy* **13** 375–404
[4] Holmstrom B and Milgrom P 1991 Multitask principal agent analyses: Incentive contracts, asset ownership, and job design *Journal of Law, Economics and Organizations* **7** 24–52
[5] Akerlof G A and Kranton R E 2010 *Identity Economics: How Our Identities Shape Our Work, Wages and Well-Being* (Princeton NJ: Princeton University Press)
[6] Ramalingam A and Rauh M T 2010 The firm as a socialization device *Management Science* **56** 2191–2206
[7] Ewing J A 1881 On the production of transient electric currents in iron and steel conductors by twisting them when magnetized or magnetizing them when twisted *Proceedings of the Royal Society of London* **33** 21–23
[8] Krasnosel'skii M A and Pokrovskii A V 1989 *Systems with Hysteresis* (Berlin: Springer)
[9] Bertotti G and Mayergoyz I D eds 2006 *The Science of Hysteresis* Vols 1-3 (Amsterdam: Elsevier Academic Press)
[10] Ford J D 1980 The occurrence of structural hysteresis in declining organisations *The Academy of Management Review* **5** 589–598
[11] Preisach F 1935 Über die magnetische Nachwirkung *Zeitschrift fur Physik* **94** 277–302
[12] Mayergoyz I D 1991 *Mathematical Models of Hysteresis* (Berlin: Springer-Verlag)
[13] Visintin A 1994 *Differential Models of Hysteresis* (Berlin: Springer)
[14] Sethna J. P., Dahmen K., Kartha S., Krumhansl J. A., Roberts B. W. and Shore J. D. 1993 Hysteresis and hierarchies: Dynamics of disorder-driven first-order phase transformations *Phys. Rev. Lett.* **70** 3347–3350
[15] Pokrovskii A and Rachinskii D 2013 Effect of positive feedback on Devil's staircase input-output relationship *Discrete and Continuous Dynamical Systems S* **6:4** 1095–1112
[16] Krejčı P, O'Kane J P, Pokrovskii A and Rachinskii D 2012 Properties of solutions to a class of differential models incorporating Preisach hysteresis operator *Physica D: Nonlinear Phenomena* **241:22** 2010–2028

IOP Conf. Series: Journal of Physics: Conf. Series **811** (2017) 012012 doi:10.1088/1742-6596/811/1/012012

On the rationale for hysteresis in economic decisions

Luis A Rios
Wharton School of Business, University of Pennsylvania, USA
E-mail: luisrios@wharton.upenn.edu

Dmitrii Rachinskii
Department of Mathematical Sciences, The University of Texas at Dallas, USA
E-mail: dmitry.rachinskiy@utdallas.edu

Rod Cross
Department of Economics, University of Strathclyde, UK
E-mail: rod.cross@strath.ac.uk

Abstract. In the social sciences there are plausible reasons to postulate that hysteresis effects are important. The available evidence, however, is predominantly at the macro level. In this paper we review the evidence regarding hysteresis in the neural processes underlying human behavior. We argue that there is a need for experimental and neuroimaging studies to fill the gap in knowledge about hysteresis processes at the micro level in the social sciences.

1. Introduction

The term hysteresis, derived from the Greek "to come behind", entered the lexicon of science to describe effects on electromagnetic fields in ferric metals that remain after a magnetizing force has been applied then removed: "these curves exhibit, in a striking manner, a persistence of previous state, such as might be caused by molecular friction the curves for the back and forth twists are irreversible, and include a wide area between them... to this action... the author now gives the name Hysteresis" ([1], p. 22).

Subsequently hysteresis effects have been identified, or postulated to exist, in a wide range of areas of inquiry, ranging from the original physics context to biology, electronics, materials science, cognitive neuroscience through to social sciences such as economics (see the entries in [2]). The mathematics of hysteresis as a general systems property has been formalised [3], showing that systems containing heterogeneous elements at

IOP Conf. Series: Journal of Physics: Conf. Series **811** (2017) 012012 doi:10.1088/1742-6596/811/1/012012

the micro level, which respond non-linearly to changes in inputs, have outputs that contain a selective, erasable memory of the non-dominated extremum values of the input innovations experienced (see [4] for a graphical exposition).

In social sciences such as economics it has been argued that hysteresis is pervasive: "the fact that the individual's continuous adjustment to changing price and income conditions changes his tastes seems so obvious that in the past economists mentioned it in passing, if at all... in order to determine the equilibrium of the consumer... we need to know... this particular hysteresis law" ([5], p. 126); see [6] for a review that stresses the distinction between ontological and epistemological hysteresis. In major areas of economic inquiry such as the effects of recessions and booms on subsequent economic activity [7, 8], the determinants of unemployment [9] and the effects of foreign exchange rate changes on international trade [10], hysteresis effects have been identified at the aggregate or macro level. Unlike many of the areas outside the social sciences in which hysteresis effects have been discovered, however, in the social sciences the evidence at the macro level has been supported more by speculation as to the behavior at the micro level that might give rise to hysteresis than by direct experimental evidence. In the original ferromagnetic context, for example, the discovery of hysteresis in the behavior of the electromagnetic field as a whole has been complemented by experimental evidence of non-linear responses by heterogeneous elements at the molecular level in the form of such as Barkhausen effects. The challenge, then, for hysteresis explanations in the social sciences is to see if tests can be conducted to see if the hysteresis in human behavior postulated to exist can be identified experimentally.

To pave the way for finer grain investigations of hysteresis in economics and the social sciences, the contribution of the present paper is to speculate as to the processes within the human brain that might give rise to hysteresis in economic and social decisions. To do this we discuss the brain memory capacity problem raised by [11] and some evidence from neuroscience that suggests the existence of hysteresis in at least some of the neural processes at work within the human brain. It is not necessarily the case that finer grain accounts of hysteresis will have greater explanatory power than accounts based on a higher level of aggregation [12], but the potential for enhanced understanding and new discoveries is there.

In the mainstream neoclassical model in economics, economic systems are characterised by equilibrium processes in which equilibria are pinned down by exogenously given consumer preferences and production relationships, and so not influenced by the paths taken towards the equilibria. This neoclassical model in economics was originally inspired by metaphors drawn from Newtonian mechanics or conservative fields of force [13], and later reformulated on the basis of axioms in which conservation and reversibility properties hold. The existence of hysteresis in neural processes would be difficult if not impossible to reconcile with this framework of analysis. Hysteresis offers a different "constitutive" metaphor [14] for understanding the behavior of economic systems. In what follows we discuss some features of what is known about neural processes that might give rise to hysteresis in economic decisions. This is best regarded as an attempt to identify pieces in the formidable jigsaw puzzle that is the human brain. These might be useful in future investigations in the burgeoning sciences of experimental economics (see [15] for a methodological review) and neuroeconomics (see [16] for a survey).

2. The von Neumann puzzle

In his posthumously published comparison of computers and human brains, von Neumann raised the question of how the storage capacity of the brain could cope with the sensory inputs experienced during a lifetime [11]. He used 10^{10} as an estimate of the number of neurons in the human brain. If it is assumed that each neuron has 10^3 connections with other neurons via axons and dendrites, and that each of the resulting synapse connections can hold one bit of information, this yields $10^{10} \times 10^3 = 10^{13}$ bits of information that the human brain can hold. The estimate of the sensory input during a lifetime is derived by assuming that all the neurons and their synapse connections can be receptors of external or internal stimuli. The estimate of a feasible receipt of stimuli at 14 per second, gives a stimuli input of 14×10^{23} bits per second. The assumed lifetime was 60 years, or about 2×10^9 seconds. The stimuli would not necessarily be retained in active parts of the brain, but, if no stimulus could be erased, the estimate of the "memory" capacity required to cope with a lifetimes stimuli is $14 \times 10^{13} \times 2 \times 10^9 = 2.8 \times 10^{23}$ bits. Thus there is a discrepancy of the order of 10^{10} bits between the brain's storage capacity and the external and internal stimuli experienced during a lifetime ([11], pp. 63–64, with the 10^3 synapse adjustment for axonal and dendritic connections from [17], p. 176).

The obvious way to reconcile the stimuli experienced during a lifetime with the brain's capacity to cope with such stimuli is to postulate that not all stimuli are stored in the brain's memory. Landauer [18] used four approaches to estimating the external stimuli that are stored during a human brain's lifetime. The first two are based on reading recall and picture recognition, and follow von Neumann's assumption that all recorded information is entered into long-term memory and is kept permanently. For the concentrated reading task the reading rate of 3 words per second was accompanied by a commitment to long-term memory of 0.4 bits per word read, giving an input to long-term memory of 1.2 bits per second. A linear accumulation over a 70-year lifetime yielded an estimate of 1.8×10^9 bits of information held in the human memory. A somewhat similar calculation based on picture recognition involved an input into long-term memory of 2.3 bits per second, cumulating to 3.4×10^9 bits over a lifetime. A third estimate allowed for an overwriting form of memory loss, giving a 1.4×10^9 bits over a lifetime estimate. A fourth estimate uses a word dictionary analogy applied to 15 domains of knowledge to produce an estimate of 0.5×10^9 lifetime bits of information ([18], p. 481, Table 1). A major limitation of the Landauer estimates is that they deal only with declarative/explicit memories and do not consider procedural/implicit memory processes within the brain.

Matters become more complicated once the distinction between short and long-term memory is introduced. An effective memory system has to not only store information but also encode and retrieve that information. This requires the brain to respond to the internal stimuli involved in encoding and retrieval operations. The working memory model ([19], Ch. 3) provides an account of how the brain's architecture intermediates between short and long-term memory. A central executive controls attention and coordinates the following: a visuo-spatial sketchpad that sets up and manipulates images from the external stimuli received; a phonological loop that is responsible for inner "speech" within the brain; and an episodic buffer that integrates information retrieval from long-term memory with the current demands of working memory. Experiments suggest that working memory capacities are quite small, subjects finding it difficult to remember sequences of numbers that contain 5 or more digits ([19], p. 44). Cowan [20] reviews the literature on the "magical number 4 in short-term memory".

Further complications arise once the assumption that long-term memories are "hard wired" is dropped. The traditional view was that short-term or working memories were labile until consolidated into long-term memory after a lapse of time, measured in hours for "synaptic consolidation" and longer for "systems consolidation" for memories that are initially hippocampus dependent ([21], 2010, p. 27). There is, however, evidence that long-term memories can also become labile during the process of retrieval. The conjecture is that process of retrieval leads to a process of reconsolidation of long-term memories [22]. This would add to the processing capacity demands on the brain, arising from the reconsolidation of long-term memories.

Finally, but by no means exhaustively, the von Neumann estimate would have to be adjusted for differences in synaptic plasticity and strength. Bartol Jr, Bromer, Kinney, Chirillo, Bourne, Harris and Sejnowski [23] found that the excitory synapses on dendritic spines of hippocampal pyramidal neurons had a minimum of 26 different synaptic strengths. In terms of the binary bits of information used in the von Neumann calculations, these estimates translate into 4.7 bits that can be stored at each synapse as synaptic strength, that is $2^{4.7} \approx 26$. Adjusting for synaptic strength differences not only raises the estimate of the memory capacity of the brain, but also the von Neumann type estimate of the stimuli that the brain's receptors could accommodate over a lifetime. In this area of research "one of the most intriguing findings is that the size of the synaptic response is not invariant, but can be altered by a variety of homo- and heterosynaptic factors such as past patterns of use or modulatory transmitters" ([24], p. 1).

3. Hysteresis in neural processes

The von Neumann puzzle was that, in the Heraclitean flux of a lifetime, the human brain is subject to external and internal stimuli that would, if all such stimuli were stored in its memory, exceed its encoding, storage and retrieval capacity. The most obvious way to resolve this puzzle is to postulate that not all stimuli received are encoded or stored in the memory. In relation to short-term memories that last for only a few minutes this would mean that not all stimuli received find a response in terms of changes in synaptic strength. In relation to long-term memories that require the cellular synthesis of new proteins, and involve short-term memory changes that are stabilized or consolidated to last for days or years ([24], p. 7), this would mean that not all changes in synaptic strength trigger the synthesis of new proteins. In this section we discuss whether thresholds of the type involved in the elemental "lazy relay" model of hysteresis could help resolve the von Neumann puzzle. It is interesting that von Neumann himself entertained this possibility: "it has been proposed to assume that the thresholds – or, more broadly stated, the stimulation criteria – for various nerve cells change with time as functions of the previous history of that cell ... thus frequent use of a nerve cell might lower its threshold, i.e. ease the requirements of its stimulation, and the like... if this were true, the memory would reside in the variability of the stimulation criteria... it is certainly a possibility, but I will not attempt to discuss it here" ([11], p. 64).

The possibility that hysteresis processes are at work in the brain, and can help resolve von Neumann's puzzle, was pursued in work conducted at the Weizmann Institute of Science in Israel in the 1960s on the physical changes that might account for the memory imprint of a pattern of external stimuli on a chemical system. Katchalsky and Neumann ([17], p. 178) argued that "the requirements of a sufficiently fast record with low expenditure of energy, and the storage of the record after the stimulus is over, severely limit the choice of possible physical mechanisms... a long retention of the memory imprint is necessary to allow the induction of further consolidation in permanent structures... it seems that this imprint could be based on equilibrium states which are a priori time-independent and stable... but any equilibrium processes are a priori excluded, since equilibrium states are independent of path, and therefore memoryless... the only simple mechanism which seems to answer all biophysical requirements is that of hysteresis".

Following the Cox, Jones, Marsh and Peacocke [25] discovery of molecular hysteresis in acid-base titration of RNA, Neumann and Katchalsky [17] studied the relationship between protonation in a synthetic polyribonucleide, poly (A) and poly (U), combination proxy for natural RNA and acid-base titration. The resulting proton binding curves are illustrated in Figure 1. The curves describing the degree of proton uptake, α, as a function of the acid-base titration, pH, exhibit the loops typical of systems with hysteresis. The α values for ascending values of pH are lower than those for descending values; and a given value of the output, α, can be associated with a range of values for the input stimulus, pH. This can give rise to a form of temporary inertia or metastability in which the output memory can coexist with a range of input stimuli. These hysteresis loops are of the type observed by Ewing [1] for electromagnetic fields in ferric metals.

The calculations underlying the von Neumann puzzle assume that sensory perception is an inevitable reflexive reaction to being exposed to a stimulus. In contrast, neuroimaging experiments suggest that "perceiving objects in our physical environment... can be formulated as a dynamic process of selecting and matching a sensory input to that predicted on the basis of higher-order representations... perceptual awareness results from this constructive process, and it is essential to stabilize percepts against continuous and often critical shifts of low-level stimulus parameters, such as contrast levels, that occur at the level of the sensory receptors" ([26], p. 660). This means that the brain does not respond to all sensory inputs, perceptual awareness only being triggered once critical or threshold values of the stimuli are reached. If there were a single critical value for a stimulus to trigger perceptual awareness, small variations around this value would involve the expenditure of energy associated with frequent changes in awareness. Instead the evidence suggests that perceptual hysteresis pertains, there being a range of values of the stimulus over which perceptual awareness will not change. This is the "lazy relay" archetypal form of hysteresis.

Figure 2 illustrates perceptual hysteresis for an experiment in which the intensity of the image of the letter k is contrasted relative to a background by slowly increasing the dot density defining the letter. The experimental subjects were asked to press a key when the letter k became clearly perceived, illustrated as the "pop out" value a in Figure 2. The image intensity was then slowly reduced from a supra-threshold level.

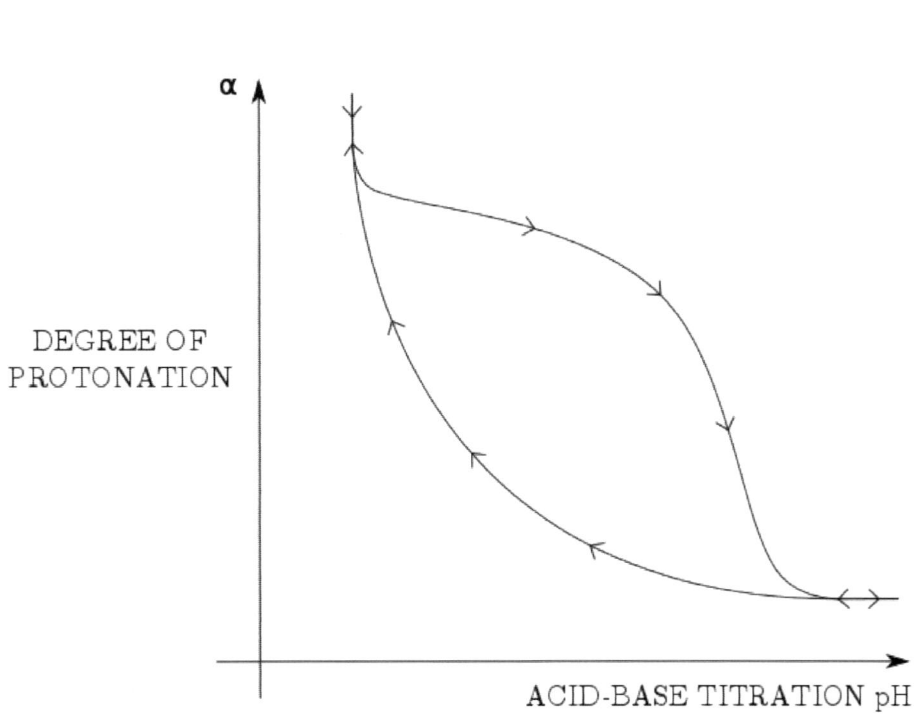

Figure 1. Hysteresis in proton binding curves. Source: Adapted from Katchalsky and Neumann (1972, p. 180, Figure 4).

Instead of the perception of k dropping out when the image intensity reached a, the "drop out" only occurred when the lower image intensity at b was reached. Hence there is a range of stimuli from b to a over which perceptual awareness does not change. As in the "lazy relay" hysteresis archetype, perceptual awareness within this range will depend on whether the percept stimulus has been approached from above or below. The neuroimaging procedures revealed that the fMRI signal changes associated with neural hysteresis were associated with the "higher" ventral lateral occipital, inferior parietal, premotor and inferior prefrontal cortex areas of the brain, and not in the early visual areas; and that the signal changes indicated greater activity when the experimental subjects were first exposed to a cycle of variation of image intensity than in subsequent repetitions of the cycle ([26], p. 659).

The existence of perceptual hysteresis might provide a clue as to how motion aftereffect illusions arise. The standard instance of this type of perceptual illusion is the waterfall effect [27]. Gaze in a concentrated way at a waterfall for about 30 seconds, and then refocus on the rockface at the side of the waterfall: the rockface appears to be moving upwards, an illusion that can last for up to a minute. This illusion may arise from the stabilization of the perception of water falling downwards by perceptual hysteresis, and the initially unsuccessful attempt by the brain to reconcile this with the

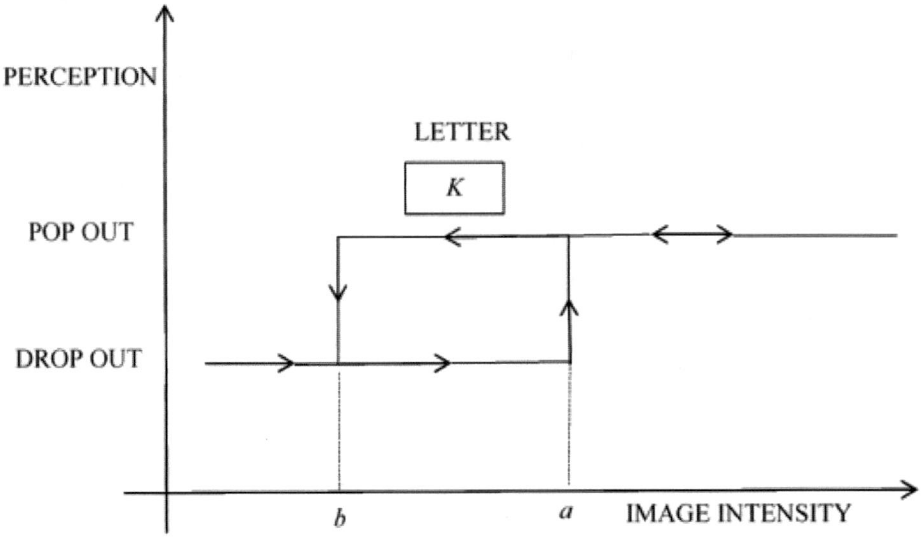

Figure 2. Perceptual hysteresis. Source: Adapted from Kleinschmidt, Büchel, Hutton, Friston and Frackowiak (2002, p. 660, Figure 1).

already stabilized perception of a stationary rockface.

Mach [28] proposed that the brain uses classification devices to cope with what would otherwise be a bewildering variety of sensory stimuli. It is now uncontroversial to postulate that prior knowledge is combined with sensory stimuli to shape perception. In cognitive neuroscience a distinction has been drawn between two effects that previous experience can have on perception: hysteresis and adaptation [29]. Hysteresis is seen as an attractive effect that sensitizes the brain to perceive the same again, whereas adaptation is a repulsive effect whereby something different is perceived. Schwiedrzik et al. [29] report fMRI evidence that the two processes map into different cortical networks: a widespread network of higher-order visual and fronto-parietal areas of the brain deal with the perceptual stabilization involved in hysteresis; whereas only early visual areas of the brain are involved in the perceptions of something different involved in adaptation. They argue that such a division of labor is efficient, or in Mach's term "economical". Hysteresis paves the way to predict what will happen next, stabilizing perception in the face of the flux of low-level stimuli which require no further processing. This allows the early visual areas of the brain to specialize on the extraction of new information from stimuli that are notably different, and so require the detailed and sensitive representations involved in adaptation. This is a Bayesian world in which hysteresis effects are the embodiment of prior probabilities about the sensory world, and changes in sensory experience represent adaptation effects that modify the likelihood functions.

4. Discussion

In the social sciences some, if not all, of the processes under study have been seen as haunted by a memory of the past such as can be found in systems with hysteresis. The evidence cited in support of such conjectures has tended to be at the macro level, with the micro processes that might give rise to hysteresis being more a matter of speculation than established fact. In this paper we have considered some evidence regarding neural processes that might help pave the way for experimental studies that would help fill this gap regarding hysteresis processes at the micro level in the social sciences. Is there perceptual hysteresis in the social sphere? Are the decisions made within organizations characterised by hysteresis? And so on.

An obvious first line of inquiry would be to conduct laboratory experiments to test for the existence of hysteresis in the decisions made by experimental subjects. A basic experiment would involve varying an incentive, such as the pay rate for a work task, upwards to discover the pay rates at which the subjects switch from a leisure to a work mode; and then varying the pay rate downwards to find the pay rates at which the subjects switch back from the work to the leisure mode. If the pay rates at which the subjects switch from leisure to work modes are significantly higher than those for switching back to leisure from work, once the work mode has been experienced, this would provide direct experimental evidence of a lazy relay hysteresis relationship between incentives and behavior. The way would then be clear for finer grain neuroimaging investigations, of the areas of the brain that are activated when switching behavior occurs, for example.

The von Neumann type estimates of brain capacity reviewed in this paper suggest that there is a marked discrepancy between the external and internal stimuli a human brain is exposed to during a lifetime and the brain's capacity to encode, store and retrieve this information when guiding behavior. It might well be the case that this discrepancy is resolved within the brain by the use of economy principles in which hysteresis plays an important role. As an anonymous referee has pointed out, hysteresis may also convey evolutionary advantages. Strategies yielding behavior that changes too frequently, or not often enough, in relation to challenges presented by the environment, are not conducive to survival. So hysteresis in the neural processes underlying behavior may have emerged by natural selection.

Acknowledgments

DR acknowledges the support of NSF through grant DMS-1413223.

References

[1] Ewing J A 1881 On the production of transient electric currents in iron and steel conductors by twisting them when magnetized or magnetizing them when twisted *Proceedings of the Royal Society of London* **33** 21–23
[2] eds G Bertotti and I D Mayergoyz 2006 *The Science of Hysteresis* (Oxford: Academic Press) vol 1
[3] Krasnosel'skii M A and Pokrovskii A V 1989 *Systems with Hysteresis* (Berlin: Springer)
[4] Mayergoyz I D 1991 *Mathematical Models of Hysteresis* (Berlin: Springer)
[5] Georgescu-Roegen N 1971 *The Entropy Law and the Economic Process* (Cambridge MA: Harvard UP)
[6] Elster J 1976 A note on hysteresis in the social sciences *Synthese* **33** 371–391

[7] Cross R, McNamara H and Pokrovskii A V 2012 Memory of recessions *Journal of Post Keynesian Economics* **34** 409–425
[8] Cross R, McNamara H, Pokrovskii A V and Rachinskii D 2008 A new paradigm for modelling hysteresis in macroeconomic flows *Physica B: Condensed Matter* **403** 231–236
[9] Darby J, Cross R and Piscitelli L 2006 Hysteresis and unemployment: a preliminary investigation The Science of Hysteresis eds G Bertotti and I D Mayergoyz (Oxford: Academic Press) vol 1 chapter 8 pp 667–699
[10] Amable B, Henry J, Lordon F and Topol R 1991 "Strong" hysteresis: an application to foreign trade *OFCE Working Paper* **9103** (Paris: OFCE)
[11] Von Neumann J 1958 *The Computer and the Brain* (New Haven: Yale UP)
[12] Jackson F and Pettit P 1992 In defence of explanatory ecumenism *Economics and Philosophy* **8** 1–21
[13] Mirowski P 1989 *More Heat than Light* (New York: Cambridge UP)
[14] Klamer A and Leonard T C 1994 So Whats an Economic Metaphor? *Natural Images in Economics* ed P Mirowski (Cambridge UP: New York) chapter 2 pp 20–51
[15] Guala F 2005 *The Methodology of Experimental Economics* (New York: Cambridge UP)
[16] Camerer C, Loewenstein G and Prelec D 2005 Neuroeconomics: How neuroscience can inform economics *Journal of Economic Literature* **43** 9–64
[17] Katchalsky A and Neumann E 1972 Hysteresis and molecular memory record *International Journal of Neuroscience* **3** 175–182
[18] Landauer T K 1986 How much do people remember? Some estimates of the quantity of learned information in long-term memory *Cognitive Science* **10** 477–493
[19] Baddeley A 2004 *Your Memory: a Users Guide* (London: Carlton Books)
[20] Cowan N 2000 The magical number 4 in short-term memory: A reconsideration of mental storage capacity *Behavioral and Brain Sciences* **24** 87–185
[21] Nader K and Einarsson E O 2010 Memory Reconsolidation: An update *Annals of the New York Academy of Sciences* **1191** 27–41
[22] Alberini C M 2011 The role of reconsolidation and the dynamic process of long-term memory and storage *Frontiers in Behavioral Neuroscience* **5** 1–10
[23] Bartol T M Jr, Bromer C, Kinney J, Chirillo M A, Bourne J N, Harris K M and Sejnowski T J 2015 Nanoconnectomic upper bound on the variability of synaptic plasticity *eLife* **4** e10778
[24] Mayford M, Siegelbaum S A and Kandel E R 2012 Synapses and memory storage *Cold Spring Harbor Perspectives in Biology* **4** a005751
[25] Cox R A, Jones A S, Marsh G E and Peacocke A R 1956 On hydrogen bonding and branching in a bacterial RNA *Biochemical Biophysics Acta* **21** 576–577
[26] Kleinschmidt A, Büchel C, Hutton C, Friston K J and Frackowiak S J 2002 The neural structures expressing perceptual hysteresis in visual letter recognition *Neuron* **34** 659–666
[27] Anstis S, Verstraten F A J and Mather G 1998 The motion aftereffect *Trends in Cognitive Sciences* **2** 111–117
[28] Mach E 1959/1897 *The Analysis of Sensations* English edition (New York: Dover Books)
[29] Schwiedrzik C M, Ruff C C, Lazar A, Leitner F C, Singer W and Melloni L 2014 Untangling perceptual memory: Hysteresis and adaptation map into separate cortical networks *Cerebral Cortex* **24** 1152–1164

IOP Conf. Series: Journal of Physics: Conf. Series **811** (2017) 012013 doi:10.1088/1742-6596/811/1/012013

Use of Prandtl-Ishlinskii hysteresis operators for Coulomb friction modeling with presliding

Michael Ruderman
University of Agder (UiA), 4604 Kristiansand, Norway
E-mail: michael.ruderman@uia.no

Dmitrii Rachinskii
University of Texas in Dallas (UTD), 800 W Campbell Road Richardson, Texas, USA
E-mail: dmitry.rachinskiy@utdallas.edu

Abstract. Prandtl-Ishlinskii stop-type hysteresis operators allow for modeling elasto-plasticity in the relative stress-strain coordinates including the saturation level of the residual constant-tension flow. This lies in direct equivalence to the force-displacement characteristics of nonlinear Coulomb friction, whose constant average value at unidirectional motion depends on the motion sign only, after the transient presliding phase at each motion reversal. In this work, we analyze and demonstrate the use of Prandtl-Ishlinskii operators for modeling the Coulomb friction with presliding phase. No viscous i.e. velocity-dependent component is considered at this stage, and the constant damping rate of the Coulomb friction is combined with the rate-independent losses of presliding hysteresis. The general case of Prandtl-Ishlinskii operator with a continuous distribution function is considered together with a finite elements case, which is useful for implementation in multiple applications. Finally, identification of parameters is addressed and discussed along with two experimental examples.

1. INTRODUCTION

Coulomb friction [1], which is well-established and often understood as a "classical" dry friction law, expands the Amontons' theory, which postulates that the tangential friction force F is μ-proportional to the normal load L. Coulomb postulated the dry friction law of the form

$$F = \mu L + C\,, \tag{1}$$

where the correction term C allowed for dependence of the friction force on the surface reactions on a microscopic (molecular) level, such as surface asperities in contact, and is independent of the normal loading and contact surface [2]. The standard notation of the Coulomb friction law for steady-state kinetic friction assumes a discontinuous dependence of the friction coefficient on the relative sliding velocity $\dot{y}$, so that

$$F = \mu L \,\mathrm{sign}(\dot{y})\,. \tag{2}$$

For constant normal loads and homogenous, time-invariant, frictional surfaces, a lumped Coulomb friction coefficient $\mu L := F_c$ is usually assigned. In order to avoid discontinuity in the Coulomb friction law (2), and therefore make it appropriate for numerical simulations and control applications, in particular at zero velocity crossing, several empirical and heuristic approaches have been proposed for capturing the so-called presliding behavior of kinetic friction (see e.g. [3, 4] for details). Examples of such approaches include the Dahl model [5], Maxwell-slip model [6], modified Maxwell-slip model [7]. Note that more comprehensive dynamic friction models, such as LuGre model [8], generic model at asperity level [9], and two-state dynamic friction model with elasto-plasticity [10], also incorporate the presliding friction behavior. The presliding friction effects have been observed in various tribological and other friction-related experiments, see e.g. [3, 11, 12, 13].

An interesting fact is that the so-called Maxwell-slip model structure [6] coincides with the Prandtl-Ishlinskii (PI) stop-type hysteresis model, for details on the latter we refer to [14, 15, 16]. This model relies on elementary Prandtl-Ishlinskii stop-type hysteresis operators, which are rheological elements well established for describing effects of elasto-plasticity. The Prandtl-Ishlinskii stop-type hysteresis operator is dual to the Prandtl-Ishlinskii play-type operator (see e.g. [16] for details) which, in turn, is well-known in mechanics and is used for describing the kinematic play, also known as backlash, in the machine elements and mechanisms. The Prandtl-

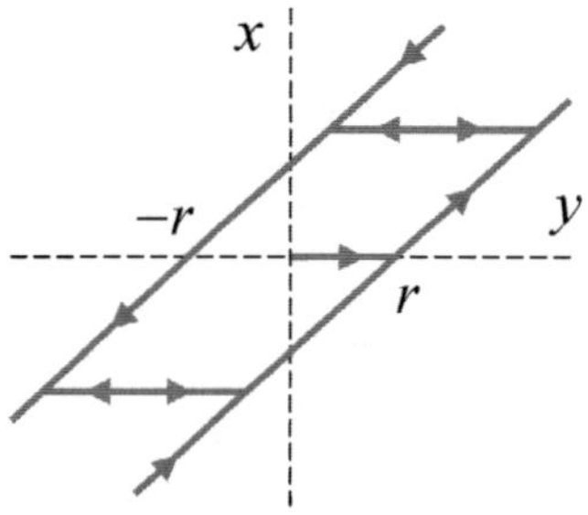

Figure 1. Prandtl-Ishlinskii play-type operator transitions.

Ishlinskii play-type operator is schematically illustrated in Figure 1. The operator provides a rate-independent map of time series y to time series x, which can be considered in various functional spaces such as $W^{1,1}$, C, BV and the space of regulated functions. This map depends on the initial state x_0 and threshold parameter $r > 0$. The latter determines the width $2r$ of the play-zone. The operator in the differential form is given by

$$\dot{x} = \begin{cases} \dot{y} & \text{if } x = y - r,\ \dot{y} \geq 0\,, \\ \dot{y} & \text{if } x = y + r,\ \dot{y} \leq 0\,, \\ 0 & \text{if } y - r < x < y + r\,, \end{cases} \qquad |x - y| \leq r \tag{3}$$

for a.e. $t \geq t_0$.

The general Prandtl-Ishlinskii model is formulated as a superposition (parallel connection) of elementary stop-type operators. A set of free parameters is used to map the distribution of elementary stop-type operators to the actual shape of hysteresis loops of the Prandtl-Ishlinskii model. Actually, it is the initial loading curve, often referred to as the "virgin curve" [17] (due to the memory-free "virgin" contacts on the frictional interface) that determines the shape of the force-displacement transitions and therefore the presliding friction hysteresis. According to Madelung's rules originally proposed in magnetism [18], and quite similar Masing's rules with origins in structural mechanics [19], measuring the initial loading curve is sufficient for a complete identification of parameters of the Prandtl-Ishlinskii hysteresis model.

The aim of this paper is to discuss how the Prandtl-Ishlinskii hysteresis operators can be efficiently used for modeling nonlinear Coulomb friction with presliding hysteresis transitions. In real-life mechanical systems to be simulated and controlled the problem of Coulomb friction discontinuity is one of the most crucial issues for proper system analysis, proof of stability, and control design. An appropriate incorporation of presliding phase into Coulomb friction law is an essential condition for the model-based analysis of system dynamics, damping properties, and synthesis of friction compensation methods. In particular, motion control problems of exact positioning type, see e.g. [20, 21], require a better perception of presliding friction behavior at the motion onset and reversals. Presliding friction effects are particularly important in such applications as rolling bearing nanopositioning stages [22], hard disk drives [23], ball-screw-driven positioning stages [24], hydraulic cylinders [25], and others.

Several strong assumptions are made here in relation to the friction behavior to be considered:

(i) No viscous (proportional to the relative velocity) friction effects are included.

(ii) The breakaway (static friction) force due to adhesion is not considered so that no velocity weakening [4], also referred to as Stribeck effect, occurs at the transients between the presliding and gross sliding regimes.

(iii) The so-called smooth sliding with the averaged constant Coulomb friction force is considered. Hence, true stick-slip dynamics on the atomic level of frictional interfaces with related Frenkel-Kontorowa and Prandtl-Tomlinson type interactions (see [26, 27] for details) are not addressed.

The paper is organized as follows. The finite elements case with a finite number of Prandtl-Ishlinskii (PI) stop-type operators and one additional discontinuous element required for the motion system to be asymptotically stable is described in Section 2. The general Prandtl-Ishlinskii operator with a continuous distribution function is discussed in Section 3. Estimation and identification of free parameters for the proposed modeling approach is discussed in Section 4, where we also present two experimental examples. Section 5 contains conclusions.

2. FINITE ELEMENTS CASE

In the finite elements case, a parallel connection of N single elasto-plastic elements with a common input is considered. Due to their superposition and the absence of coupling between the elements, the total output friction force is given by simple superposition

$$F(y) = \sum_{j=1}^{N} F_j \, . \tag{4}$$

Note that the input argument in (4), which is the same for each single elasto-plastic element, is the relative displacement y and not the relative velocity $\dot{y}$, as it is assumed in the general case of kinetic friction addressed in Section 1. However, a differential form of PI operators with $\dot{y}$ used as the input is also available, see e.g. in [28]. Recall that each elasto-plastic element j representing a stop-type PI operator can be described by

$$F_j = k_j f_j, \qquad f_j = y - x_j(y, r_j) \, , \tag{5}$$

where k_j is the stiffness coefficient, f_j is the elastic deformation, and x_j is the output state of the corresponding play-type operator illustrated in Figure 1. The free parameters are the individual break-away force level c_j and presliding stiffness k_j of each elementary elasto-plastic element. These two parameters are related through the threshold value of the play-type PI operator, i.e.

$$r_j = c_j / k_j \, . \tag{6}$$

One can recognize that both parameters control the individual transition between presliding and sliding and therefore determine piecewise linear segments of the force-displacement presliding hysteresis curve. An example of such curve with three elasto-plastic elements is schematically illustrated in Figure 2. The shadowed regions characterize the presliding range. It is important

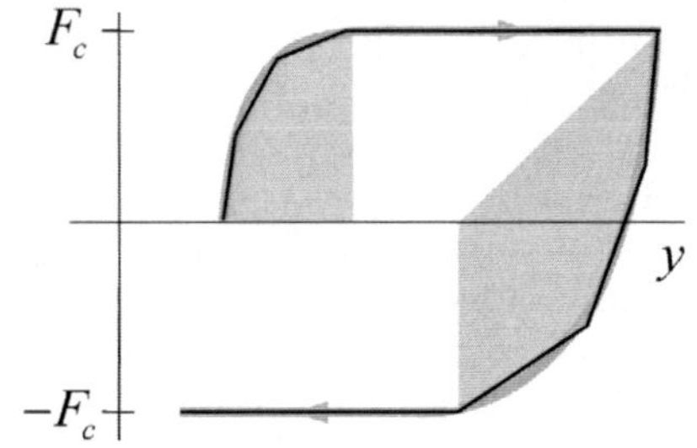

Figure 2. Presliding friction behavior and its piecewise linear approximation.

to note that after each motion reversal and at the onset of relative motion, each elasto-plastic element in the structure is first sticking before passing into the sliding. Therefore, the initial presliding stiffness after each motion reversal, which occurs at a relative position $y(t_r)$, is equal to the cumulative stiffness of all elasto-plastic elements so that

$$\frac{\partial F(t_r^+)}{\partial y(t_r^+)}\Big|_{y=y(t_r)} = \sum_{j=1}^{N} k_j \,. \tag{7}$$

It has been shown in [28] that for an unforced motion system (with friction force F) to be globally asymptotically stable in the Lyapunov sense, i.e. to have the global zero motion equilibrium, at least one elasto-plastic element should have $k_j \rightarrow \infty$ (see Corollary 2 in [28]). In other words, after each motion reversal at least one elasto-plastic element is immediately in sliding. This corresponds to a constant Coulomb-type friction force with discontinuity at the velocity zero-crossing. Recall that the overall Coulomb friction force with discontinuity, i.e. $F \sim \mathrm{sign}(\dot{y})$, constitutes the limit case of presliding hysteresis curves at motion reversals, see [29] for details.

In order to illustrate the initial stiffness behavior of presliding friction hysteresis, consider a finite elements case with $N = 3$ stop operators with $c_{1-3} = [1, 1, 1]$. We set $k_{2,3} = [10, 4]$, and vary the stiffness of the first element by an order of magnitude, $k_1 = \{500, 50, 5\}$. Friction-displacement hysteresis curves obtained from the same input sequence are shown in Figure 3.

In order to make the finite elements friction modeling globally asymptotically stable, equation (4) should be modified in such a way that at least one element does not exhibit an elastic phase and passes into sliding immediately after each motion reversal. This can be achieved by adding the Coulomb friction law with discontinuity, thus yielding

$$F(y, \dot{y}) = c_1 \,\mathrm{sign}(\dot{y}) + \sum_{j=2}^{N} F_j \,. \tag{8}$$

It is obvious that $0 < c_1 < F_c$ should hold while

$$\sum_{j=1}^{N} c_j = F_c \,, \tag{9}$$

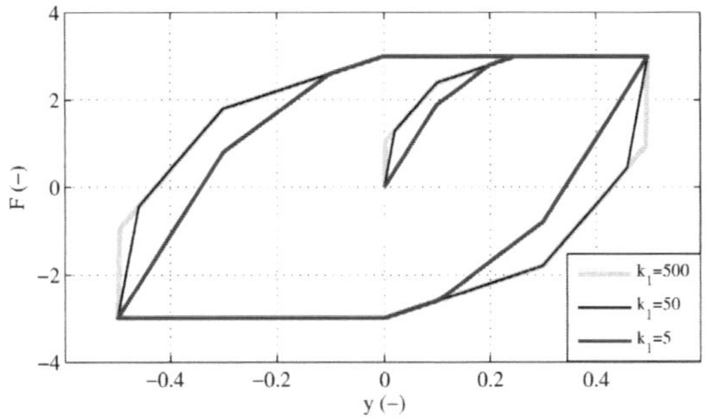

Figure 3. Presliding friction hysteresis curves for a varying stiffness of one elasto-plastic element within the model with 3 PI stop-type operators.

where F_c is the Coulomb friction coefficient, which determines the maximal friction amplitude at unidirectional motion. Several important properties of the presliding friction behavior can be derived for model (8):

- For $c_1 > 0$, the total friction behavior has infinite stiffness at each motion reversal, independent of presliding or sliding state in the relative force-displacement coordinates. This is in accord with stiction friction phenomenon, for which a certain amount of external force is first required to move an idle motion system with friction from zero equilibrium.
- If the elementary stiffnesses satisfy $k_j \to \infty$ for all $j \in [2, \ldots, N]$, then the total friction force satisfies $F(y, \dot{y}) \to F_c \operatorname{sign}(\dot{y})$, which corresponds to the "classical" Coulomb friction law and represents the limit case of presliding hysteresis friction.
- With $N \to \infty$, the friction force equation (8) can be transformed into the general PI model considered in the next section.

3. GENERAL CASE

A more general Maxwell-slip friction model includes an infinite number of stop-type PI operators parameterized by the threshold r, which ranges over an interval of values $0 \leq r \leq R$. Using this formalism, the friction force can be modeled by the following counterpart of relation (8):

$$F(y, \dot{y}) = c_1 \operatorname{sign}(\dot{y}) + \int_0^R f(y, r)\, d\mu(r) \tag{10}$$

with the normalized force $f(r)$ (or, equivalently, elastic deformation) given by

$$f(y, r) = y - x(y, r)\,, \tag{11}$$

where $x(y, r)$ is the output state of the play-type PI operator, which has threshold r (cf. Figure 1). The term $c_1 \operatorname{sign}(\dot{y})$ in (10) corresponds again to the element with infinite stiffness. The Riemann-Stieltjes integral with a distribution function μ in (10) is a PI operator of general type [14].

In particular, with a piecewise constant distribution

$$\mu(r) = \sum_j^N k_j H(r - r_j),$$

where H is the Heaviside step function, formula (10) with $R > \max r_j$ reduces to the finite elements case (8).

If the distribution function μ is continuous, then the unforced motion system with friction (10) is globally asymptotically stable even in the case when $c_1 = 0$, i.e. stiffness is uniformly bounded for all elements (see [28]). This is because for any motion, elements with a sufficiently small threshold r are always in sliding, hence any motion dissipates energy due to friction. This can be compared to the finite elements case (8), in which sufficiently small oscillations do not dissipate energy because all the elements are in pre-sliding mode, and the system effectively behaves as a harmonic oscillator. If $c_1 = 0$, then the motion system with a continuous distribution function μ will produce infinitely many damped oscillations of decreasing vanishing amplitude. On the other hand, if $c_1 > 0$ (i.e. there is an element with infinite stiffness), then the motion comes to a complete stop after a finite number of oscillations both in cases of discrete-valued and continuous μ distribution.

Assuming the lumped unity mass (i.e. $m = 1$), the equation of free motion with friction (10) has the form

$$\ddot{y}(t) + c_1 \operatorname{sign}\big(\dot{y}(t)\big) + \int_0^R f\big(y(t), r)\big)\, d\mu(r) = 0\,, \tag{12}$$

and the total energy of this system is

$$E = \frac{1}{2}\Big(\dot{y}^2 + \int_0^R \big(f(y,r)\big)^2\, d\mu(r)\Big)\,.$$

Note that the maximal elastic force associated with the integral term in (12) equals

$$F_m = \int_0^R r\, d\mu(r).$$

Clearly, if the stiction friction level exceeds this force,

$$c_1 > F_m,$$

then any free motion of the mass m will proceed in the direction of the initial momentum until, after a finite time, the mass will come to a complete stop because the static friction will balance the elastic force at the stopping point.

Now, assuming that the opposite inequality holds, $c_1 < F_m$, and the initial momentum (energy) is sufficiently large, let us consider the conditions, which ensure that the trajectory makes one turning point before the motion comes to a complete stop. Without loss of generality, let us suppose that the initial momentum is negative and place the origin, $y = 0$, at the first turning point of the motion. Then, the second turning point satisfies $y_e > 0$. If the initial momentum is sufficiently large, then the elastic force at the first turning point is at its maximal value F_m, and the equation of motion between the first turning moment t_1 when $y(t_1) = 0$ and the second turning moment t_2 when $y(t_2) = y_e$ reads

$$\ddot{y}(t) + c_1 + \Phi\big(y(t)\big) = 0 \tag{13}$$

with the elastic force given by

$$\Phi(y) = \int_0^{y/2} r\, d\mu(r) + \int_{y/2}^R (y - r) d\mu(r)\,, \tag{14}$$

where y increases from $y(t_1) = 0$ to $y(t_2) = y_e$ and $\dot{y}(t_1) = \dot{y}(t_2) = 0$. Multiplying (13) by $\dot{y}$ and integrating the resulting equation from t_1 to t_2 yields the following equation for the second turning point y_e:

$$c_1 y_e + \int_0^{y_e} \Phi(y)\, dy = 0\,. \tag{15}$$

The motion comes to a complete stop at this point if the stiction friction level exceeds the elastic force, that is if

$$c_1 > \Phi(y_e). \tag{16}$$

Thus, the free motion with a sufficiently high initial momentum exhibits exactly one reversal (turning point) before coming to a complete stop if condition (16) holds for the smallest positive root y_e of equation (15) with the function Φ defined by (14). Continuing the same line of argument, one obtains similar conditions, which ensure that the motion stops after making exactly n reversals.

4. IDENTIFICATION OF PARAMETERS

In most cases, the identification of parameters of the friction model is done on a real motion system, for which both a moving mass (inertia) and frictional contact interface constitute the given physical matters. Assuming a general driving input force u, which can be the control value, and the lumped mass m, the motion dynamics with friction (8) is described by

$$m\ddot{y}(t) + c_1 \operatorname{sign}\big(\dot{y}(t)\big) + \sum_{j=2}^{N} k_j\big(y(t) - x_j(y, r_j)\big) = u(t) \tag{17}$$

(cf. (12)). Assuming that the control value $u(t)$, relative displacement $y(t)$ and at least its first time derivative are the measurable quantities, the model identification reduces to finding of the best-fit of the free parameters (friction model and mass) based on the generated and recorded experimental data. Here the model identification errors can be built based on either motion or force quantities, i.e. forward or inverse model, while the other signals constitute the available regressors for identification. All the residual parameters, with the exception of thresholds of stop-type operators, contribute linearly to (17), makes the identification procedure straightforward. According to [30] and [31], both linear regression (LS) and nonlinear regression (NLS) approaches are well suited for indentifying the Maxwell-slip-type structure, which is equivalent to the PI model. For the LS approach, a pre-assignment of elements' threshold values (equivalent to the break-away force level) is however required, since these are not linear parameters according to from (17). Various numerical routines are available for solving either LS or NLS problem related to (17). Here we abstain from presenting details of identification algorithms and instead refer to the fundamental literature on system identification, e.g. [32].

Two experimental tests, both aimed at comparing the measured and modeled motion response in vicinity of an equilibrium state, i.e. full motion stop, are presented below. We note that before the full stop the moving mass can exhibit at least one detectable reversal. This effect is due to a finite non-zero inertia, which is antagonized by presliding friction with elasto-plastic interactions on the contact interface. The targeted motion phase represents an unforced system dynamics (equation (17) with zero right-hand-side) with non-zero initial conditions which are detectable from the available measurements. This operation regime is particularly suitable for exposing the presliding friction and contact stiffness at motion reversals.

In both experimental tests shown below, the identified parameters of the model with four elasto-plastic elements ($N = 5$) are m, F_c, c_1, k_2, k_3, k_4, and k_5. For the sake of simplicity and in order to avoid local minima during the curves fit, the break-away force level of the single stop-type elements are assumed to have one and the same value $c_{2-5} = (F_c - c_1)/4$. This is a reasonable pre-assignment for mapping the initial loading curve when the number of elements equals to or exceeds 3 (for more details see the identification results in [31, 33]).

Experimental test I

The first experimental test, the same as reported in [28, 29], represents a steady-state motion with the constant velocity at constant (control) input force at the beginning. Afterwards, the control input is switched-off and the relative motion decelerates to zero being driven by the counteracting friction only. The control signal and the measured relative velocity are shown in Figure 4. An almost linear slope of the decreasing relative velocity points to the predominance of the constant Coulomb friction force for which the unforced motion dynamics is $\ddot{y} = -(F_c/m)\,\mathrm{sign}(\dot{y})$. After the first velocity zero-crossing, a low negative overshoot occurs, which is the motion reversal followed by a low-velocity displacement (at time between 0.09 sec and 1.13 sec) before the final motion stop. Note that the depicted relative velocity is the discrete-time derivative of the recorded incremental encoder signal. Therefore the pulsed pattern does not imply any measurement noise but the actual low velocity, which is inherently subject to the discretization and quantization effects. The same can be said about the measured velocity pattern during the decreasing slope related to the constant deceleration. The identified model

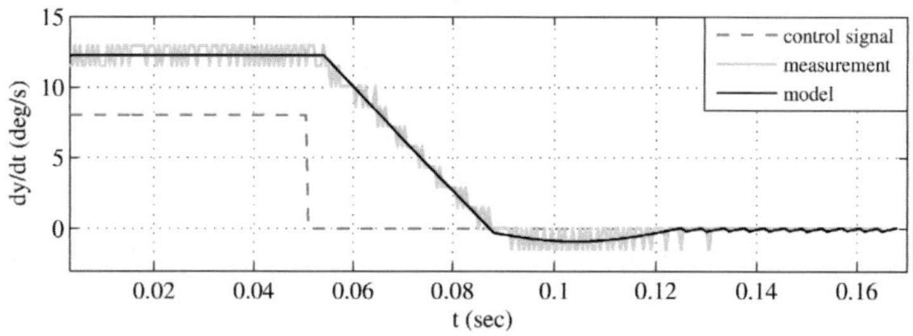

Figure 4. Experimental test I (measurement and model): velocity response in a vicinity of the motion stop after switching off the constant control force.

response is shown in Figure 4 over the measurement. The curve's best-fit exhibits, however, some residual micro-oscillations related to the numerical computation of (17), since the sign-discontinuity is incorporated.

Experimental test II

In the second experimental test, the measured data are taken from the free (unforced) system, which has been first excited by an input impulse. The data that we use is close to the motion stop and discards the transients, which can include the non-modeled high-frequency components coming from the unaccounted flexibilities in the system. The measured velocity response under consideration is depicted in Figure 5(a). Note that the experimental motion system is without gearing mechanisms, in contrast to Experimental test I, and is equipped with an angular encoder of higher resolution. This becomes apparent when comparing the increments of the measured velocity patterns in Figures 4 and 5(a). The negative velocity pattern after the first zero-crossing is more accurately captured. This provides a more detailed signature of presliding friction after the motion reversal. The details about experimental setup, from which the measured data has been obtained, can be found in [34]. The fitted model response shown in Figure 5(a) demonstrates an accurate match with the measured data. The accuracy of the fitted model becomes further apparent when analyzing the corresponding phase portrait depicted in Figure 5(b). Here the model-based prediction of relative displacement carries weight in order to decrease

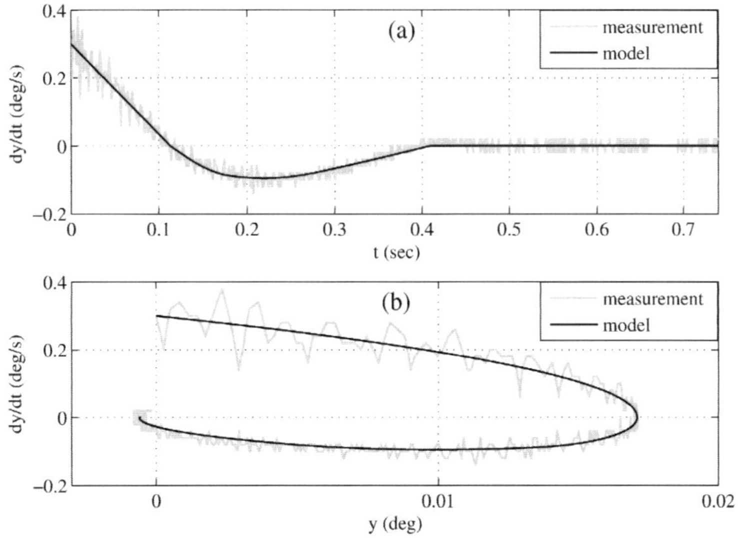

Figure 5. Experimental test II (measurement and model): (a) velocity response in vicinity the motion stop; (b) displacement-velocity phase portrait.

an integral error. From both the measured and modeled phase portraits, a marginally visible second motion reversal can be also recognized. However, this apparently lies beyond the available measurement accuracies and is not analyzed here.

5. CONCLUSIONS

In this paper, we have explored how the Prandtl-Ishlinskii stop-type hysteresis model, which is equivalent to the Maxwell-slip structure, can be used for describing the kinetic Coulomb friction including the presliding phase. The latter is characterized by the rate-independent force. The so-called initial loading ("virgin") curves determine the shape of hysteresis loops and damping characteristics of the unforced motion before the final motion stop. Finite elements and general continuous distributions have been addressed, and the parameters identification has been demonstrated with two experimental examples. We have identified conditions which ensure that the motion stops after a given number of reversals. It would be interesting to consider a counterpart of equation (17) with interacting stop nonlinearities, for example using the techniques developed in [35, 36, 37, 38].

Acknowledgments

D. Rachinskii acknowledges the support of NSF through grant DMS-1413223.

References

[1] de Coulomb C A 1785 *Memoire de Mathematique et de Physics de l'academie Royal* 161–342
[2] Awrejcewicz J and Olejnik P 2005 *Applied Mechanics Reviews* **58** 389–411
[3] Armstrong B, Dupont P and De Wit C C 1994 *Automatica* **30** 1083–1138
[4] Al-Bender F and Swevers J 2008 *IEEE Control Systems Magazine* **28** 64
[5] Dahl P R 1968 A solid friction model TOR 158(3107-18) The Aerospace Corporation, El Segundo
[6] Lazan B J 1968 *Damping of materials and members in structural mechanics* vol 214 (Pergamon press Oxford)
[7] Ruderman M and Bertram T 2011 *IFAC Proceedings Volumes* **44** 10764–10769 18th IFAC World Congress
[8] De Wit C C, Olsson H, Aström K J and Lischinsky P 1995 *IEEE Trans. on automatic control* **40** 419–425

[9] Al-Bender F, Lampaert V and Swevers J 2004 *Tribology Letters* **16** 81–93
[10] Ruderman M and Bertram T 2013 *Mechanical Systems and Signal Processing* **39** 316–332
[11] Koizumi T and Shibazaki H 1984 *Wear* **93** 281–290
[12] Bowden F P and Tabor D 2001 *The friction and lubrication of solids* vol 1 (Oxford university press)
[13] Lampaert V, Al-Bender F and Swevers J 2004 *Tribology Letters* **16** 95–105
[14] Krasnosel'skii M and Pokrovskii A 1989 *Systems with Hysteresis* (Springer)
[15] Visintin A 1994 *Differential models of hysteresis* (Springer)
[16] Krejci P 1996 *Hysteresis, Convexity and Dissipation in Hyperbolic Equations* (Tokyo: Gattötoscho)
[17] Al-Bender F, Symens W, Swevers J and Van Brussel H 2004 *International journal of non-linear mechanics* **39** 1721–1735
[18] Madelung E 1905 *Annalen der Physik* **322** 861–890
[19] Masing G and Mauksch W 1925 *Wissenschaftliche Veröffentlichungen aus dem Siemens-Konzern* 244–256
[20] Ruderman M and Iwasaki M 2016 *IEEJ Journal of Industry Applications* **5** 61–68
[21] Ruderman M 2016 *IFAC-PapersOnLine* **49** 82–86
[22] Dong X, Yoon D and Okwudire C E 2017 *Precision Engineering* **47** 375–388
[23] Yan T and Lin R 2003 *IEEE Transactions on Magnetics* **39** 1064–1069
[24] Zhu H and Fujimoto H 2015 *IEEE/ASME transactions on mechatronics* **20** 956–966
[25] Tran X B, Hafizah N and Yanada H 2012 *Mechatronics* **22** 65–75
[26] Urbakh M, Klafter J, Gourdon D and Israelachvili J 2004 *Nature* **430** 525–528
[27] Popov V and Gray J 2012 *ZAMM - J. of Applied Math. and Mech.* **92** 683–708
[28] Ruderman M 2015 *Mechatronics* **30** 225–230
[29] Ruderman M and Iwasaki M 2016 *Journal of Physics: Conf. Series* **727** 012014
[30] Al-Bender F, Lampaert V and Swevers J 2004 *Chaos* **14** 446
[31] Rizos D and Fassois S 2009 *IEEE Transactions on Control Systems Technology* **17** 153–160
[32] Ljung L 1999 *System Identification: Theory for the User* 2nd ed (Prentice Hall)
[33] Al-Bender F, Lampaert V, Fassois S, Rizos D, Worden K, Engster D, Hornstein A and Parlitz U 2003 *Nonlinear Dynamics of Production Systems* 349–367
[34] Ruderman M, Aranovskii S, Bobtsov A and Bertram T 2012 *Automation and Remote Control* **73** 1604–1615
[35] Krejci P, Lamba H, Melnik S and Rachinskii D 2014 *Phys. Rev. E* **90** 032822
[36] Krejci P, Lamba H, Melnik S and Rachinskii D 2015 *Discrete and Continuous Dynamical Systems B* **20** 2949–2965
[37] Krejci P, O'Kane P, Pokrovskii A and Rachinskii D 2012 *Physica D* **241** 2010–2028
[38] Krejci P, Lamba H, Antunes G and Rachinskii D 2016 *Mathematica Bohemica* **141** 261–286

Meshless methods for 'gas - evaporating droplet' flow modelling

Oyuna Rybdylova, Sergei S Sazhin

The Sir Harry Ricardo Laboratories, Advanced Engineering Centre,
School of Computing, Engineering and Mathematics, University of Brighton, UK

E-mail: O.Rybdylova@brighton.ac.uk; S.Sazhin@brighton.ac.uk

Abstract. The main ideas of simulation of two-phase flows, based on a combination of the conventional Lagrangian or fully Lagrangian (Osiptsov) approaches for the dispersed phase and the mesh-free vortex and thermal-blob methods for the carrier phase, are summarised. In the approach based on a combination of the fully Lagrangian approach for the dispersed phase and the vortex blob methods for the carrier phase the problem of calculation of all parameters in both phases (including particle concentration) is reduced to the solution of a high-order system of ordinary differential equations, describing transient processes in both carrier and dispersed phases. It contrast to this approach, in the approach based on a combination of the conventional Lagrangian approaches for the dispersed phase and the vortex and thermal-blob methods for the carrier phase the non-isothermal effects in the two-phase flow were taken into account. The one-way coupled, two-fluid approach was used in the analysis. The gas velocity field was restored using the Biot-Savart integral. Both these approaches were applied to modelling of two processes: the time evolution of a two-phase Lamb vortex and the development of an impulse two-phase jet. Various flow patterns were obtained in the calculations, depending on the initial droplet size.

1. Introduction

Two-phase flows are widely observed in engineering and environmental conditions (e.g. [1]). In such flows, the admixture sometimes forms high concentration regions with complex structures [1]. The Eulerian approaches cannot describe such regions with reasonable accuracy, since these approaches are based on the assumption of single-valued fields of the particle/droplet concentration and velocities. As demonstrated in [2], the only approach capable of calculating the droplet concentration field, without using excessive computer power, is the one suggested by Osiptsov [3]. The latter approach is commonly known as the Osiptsov method or approach.

Various meshless methods for the carrier phase flows have been developed and have proved to be efficient tools for investigating complex single-phase flows both with primitive and vorticity-velocity variables (e.g. [4]). Lebedeva et al. [5] developed and tested a method combining the viscous-vortex method for the carrier phase and Osiptsov's approach [3] for particles/droplets. This approach combined the advantages of both the viscous-vortex and Osiptsov methods.

The approaches mentioned above were focused on hydrodynamic aspects of particle-laden flows. However, in many engineering applications, including automotive applications [6], the effects of heat and mass transfer are significant. In [7] some aspects of the approach described in [5] were generalised to take into account some of these effects. The phase transition on the droplet surface was described using a simple model based on the assumption that the heat flux reaching the droplet is spent on its evaporation (cf. a similar assumption used for qualitative engineering analysis of droplet evaporation in multiphase flows ([6, 8]).

The aim of this paper is to present a summary of the models developed and used in the previous papers [5, 7] (Section 2) and the main results obtained (Section 3). The main results of this paper are summarised in Section 4. The publication of this mini-review can be justified by the fact that the original papers were published in engineering journals which are almost unknown to the mathematical community. At the same time, these papers are focused primarily on the description of new mathematical tools and their engineering applications, which are expected to be relevant to mathematical research in this field.

2. Models

In what follows, the main ideas of the models developed and used for 2D plane flows (Cartesian coordinates) in [5, 7] are briefly summarised.

2.1. Fully Lagrangian (Osiptsov) method

In the fully Lagrangian (Osiptsov) approach [3] the dispersed phase number density is inferred from the solutions to the following system of ordinary differential equations along chosen droplet trajectories:

$$n_s\,|J| = n_{s0} \tag{1}$$

$$\frac{\partial \mathbf{x}_s}{\partial t} = \mathbf{v}_s, \quad \frac{\partial \mathbf{v}_s}{\partial t} = \beta(\mathbf{v} - \mathbf{v}_s)\chi_d + \frac{1}{\mathrm{Fr}^2}\mathbf{e}_g, \tag{2}$$

$$\frac{\partial J_{ij}}{\partial t} = q_{ij}, \tag{3}$$

$$\frac{\partial q_{ij}}{\partial t} = \beta\left(\frac{\partial v_i}{\partial x}J_{xj} + \frac{\partial v_i}{\partial y}J_{yj} - q_{ij}\right)\chi_d + \beta\left(v_i - v_{si}\right)\frac{\partial \chi_d}{\partial x_{j0}}, \tag{4}$$

$$\frac{\partial \chi_d}{\partial x_{j0}} = \frac{1}{9}\frac{\mathrm{Re}_{s0}}{\mathrm{Re}_s^{1/3}}\frac{1}{|\mathbf{v} - \mathbf{v}_s|}\left((u - u_s)\left(\frac{\partial u}{\partial x}J_{xj} + \frac{\partial u}{\partial y}J_{yj} - q_{xj}\right) + \right.$$
$$\left. + (v - v_s)\left(\frac{\partial v}{\partial x}J_{xj} + \frac{\partial v}{\partial y}J_{yj} - q_{yj}\right)\right),$$

where

$$J_{ij} = \frac{\partial x_i}{\partial x_{j0}}, \quad q_{ij} = \frac{\partial v_{si}}{\partial x_{j0}}, \quad \chi_d = 1 + \mathrm{Re}_s^{2/3}/6,$$

$$\beta = \frac{6\pi\sigma\mu R_0^2}{m\Gamma_0}, \quad \mathrm{Fr} = \frac{\Gamma_0}{R_0\sqrt{gR_0}}, \quad \mathrm{Re}_s = \mathrm{Re}_{s0}\,|\mathbf{v} - \mathbf{v}_s|, \quad \mathrm{Re}_{s0} = \frac{2\sigma\Gamma_0}{R_0\nu},$$

indices i and j take values of x or y in the Cartesian coordinate system; $\mathbf{e}_g$ is the unit vector along the direction of the gravity force; x_{s0}, y_{s0} are the Lagrangian variables (the coordinates of initial particle positions); J is the Jacobian of the transformation from the Eulerian to the Lagrangian coordinates.

Equation (1) is the continuity equation rewritten in the Lagrangian variables. Equations (2) are momentum balance equations along chosen particle trajectories. Equations (3) and (4) are additional equations to calculate the Jacobian components; they are derived from Equations (2) by differentiation with respect to x_{s0} and y_{s0}. These equations are solved subject to standard initial conditions.

These equations were presented for plane geometry. Their generalisation to cylindrical geometry is straightforward (see [3] for the details).

2.2. *Viscous-vortex and thermal-blobs*

In the viscous-vortex and thermal-blob methods the dimensionless carrier phase equations are written in the form [7]:

$$\frac{\partial \omega}{\partial t} + \mathrm{div}\left(\omega \mathbf{v}\right) = \frac{1}{\mathrm{Re}} \Delta \omega, \tag{5}$$

$$\frac{\partial T}{\partial t} + \mathrm{div}(T\mathbf{v}) = \frac{\gamma}{\mathrm{Re}\,\mathrm{Pr}} \Delta T, \tag{6}$$

where $\omega = \nabla \times \mathbf{v}$ is the vorticity; $\mathrm{Re} = \rho LU/\mu$ and $\gamma = c_p/c_v$ are the Reynolds number and the specific heat ratio; ρ, μ and U are density, dynamic viscosity and velocity of the carrier phase. The following non-dimensional parameters were used:

$$\mathbf{r}\left(\mathbf{r}_s\right) = \frac{\mathbf{r}^*\left(\mathbf{r}_s^*\right)}{L}, \ \mathbf{v}\left(\mathbf{v}_s\right) = \frac{\mathbf{v}^*\left(\mathbf{v}_s^*\right)}{U}, T\left(T_s\right) = \frac{T_\infty - T^*\left(T_s^*\right)}{T_\infty - T_{ref}}, \ \sigma = \frac{\sigma^*}{\sigma_{ref}},$$

$$\mathbf{r} = \left(x, y\right), \mathbf{r}_s = \left(x_s, y_s\right), \mathbf{v} = \left(u, v\right), \mathbf{v}_s = \left(u_s, v_s\right),$$

where L and U are the length and velocity scales of the problem; T_{ref} and T_∞ are the gas dimensional reference temperature (e.g. jet temperature at the inlet) and the temperature in the far field, respectively, σ_{ref} is the droplet radius scale (e.g. initial droplet radius in the case of initially monodisperse droplets); the so-defined dimensionless temperature takes values between 0 and 1, with 0 corresponding to the temperature in the far field.

Equation (5) is the vorticity transport equation which follows from the Navier-Stokes equations for an incompressible fluid. Equation (6) is the transient heat conduction equation.

Introducing the vortex and thermal diffusion velocities, $\mathbf{v}_{dv}$ and $\mathbf{v}_{dT}$, Equations (5) and (6) can be presented in the divergence forms:

$$\frac{\partial \omega}{\partial t} + \mathrm{div}\left(\omega\left(\mathbf{v} + \mathbf{v}_{dv}\right)\right) = 0, \tag{7}$$

$$\frac{\partial T}{\partial t} + \mathrm{div}\left(T\left(\mathbf{v} + \mathbf{v}_{dT}\right)\right) = 0, \tag{8}$$

where

$$\mathbf{v}_{dv} = -\frac{1}{\mathrm{Re}}\frac{\nabla\omega}{\omega}, \ \mathbf{v}_{dT} = -\frac{\gamma}{\mathrm{Re}\,\mathrm{Pr}}\frac{\nabla T}{T}.$$

This allows tracking of viscous-vortex and thermal-blobs moving with velocities $v + v_{dv}$ and $v + v_{dT}$, respectively, by solving ODEs subject to corresponding initial conditions. Then vorticity and temperature fields are calculated using Equations (7) and (8).

For calculation of vorticity, the domain with a non-zero vorticity is discretised into N elements, with the area of the i-th element equal to Δ_{vi}:

$$\omega\left(\mathbf{r}, t\right) \approx \sum_{i=1}^{N} \Gamma_i \zeta_{\varepsilon_i}\left(\mathbf{r} - \mathbf{r}_{vi}\left(t\right)\right), \tag{9}$$

$$\Gamma_i \approx \omega_0\left(\mathbf{r}_{vi}\left(t_0\right)\right)\Delta_{vi} = \mathrm{const}, \tag{10}$$

where $\zeta_{\varepsilon_i}\left(\mathbf{r}\right)$ are the so-called cut-off functions. The elements of discretisation are called blobs.

Similarly, the equations for M thermal blobs take the form:

$$T\left(\mathbf{r}, t\right) \approx \sum_{i=1}^{M} \Theta_i \zeta_{\varepsilon_i}\left(\mathbf{r} - \mathbf{r}_{Ti}\left(t\right)\right), \tag{11}$$

$$\Theta_i = T_0\left(\mathbf{r}_{Ti}\left(t_0\right)\right)\Delta_{Ti} = \mathrm{const}, \tag{12}$$

where Θ_i and $\mathbf{r}_{Ti}$ are the strength and position of the i-th thermal blob.

Once the vorticity field is calculated, then the velocity field can be restored using the Biot-Savart integral.

3. Results

In this section, some results obtained using a combined fully Lagrangian and viscous-vortex blob approach (originally reported in [5]) and a combined conventional Lagrangian, viscous-vortex blob and thermal-blob approach (originally reported in [7]) are briefly summarised.

3.1. A combined fully Lagrangian and viscous-vortex blob approach

The approach based on a combination of the fully Lagrangian method for the dispersed phase and the mesh-free viscous-vortex blob method for the carrier phase was suggested and tested in [5]. In this section the main findings reported in the latter paper are briefly summarised.

This approach was applied to modelling of two processes: the time evolution of a two-phase Lamb vortex and the development of an impulse two-phase jet. These examples involve the formation of local zones of particle accumulation and regions of multiple intersection of particle trajectories. These features of the flow cannot be simulated using the conventional Eulerian and Lagrangian methods described in the literature.

3.1.1. Isothermal Lamb vortex flow The distribution of the vorticity and the azimuthal velocity of the Lamb vortex flow are described by the following equations (r is the distance from the vortex centre):

$$\omega = \frac{\mathrm{Re}}{4\pi t}\exp\left(-\mathrm{Re}\frac{r^2}{4t}\right), \tag{13}$$

$$v_\varphi = \frac{1}{4\pi}\left[1-\exp\left(-\mathrm{Re}\frac{r^2}{4t}\right)\right], \tag{14}$$

where $\mathrm{Re} = \Gamma_0/\nu_g$, Γ_0 and ν_g are the circulation of the Lamb vortex and gas kinematic viscosity, respectively. These expressions are presented in dimensionless form, where the scales for length, velocity, time, and vorticity are taken equal to the particle velocity relaxation length $\ell_\tau = (m\Gamma_0/(6\pi\sigma\mu_g))^{1/2}$, $U = \Gamma_0/\ell_\tau$, $T = \ell_\tau^2/\Gamma_0$ and $W = \Gamma_0/\ell_\tau^2$, respectively. These expressions were used in [5] for verification of the numerical algorithm for calculation of vorticity and velocity for the carrier phase.

It was shown in [5] that the vorticity value found when the number of vortex blobs is of the order of one thousand, and the value found from the exact solution are very close. This verifies the method of calculating the carrier phase parameters, described earlier.

The dynamics of the dispersed phase are described in [5] by the following system of equations:

$$\frac{\partial \mathbf{v}_s(t,\mathbf{r}_{s0})}{\partial t} = a\left(\mathbf{v}-\mathbf{v}_s\right) + b\frac{\mathrm{d}\mathbf{v}}{\mathrm{d}t} + c\mathbf{e}_g, \tag{15}$$

$$\frac{\partial \vartheta_{pq}}{\partial t} = a\left(D_q\left[(\mathbf{v})\right] - \vartheta_{pq}\right) + bD_q\left[\frac{\mathrm{d}(\mathbf{v})_p}{\mathrm{d}t}\right], \tag{16}$$

where,

$$D_q = J_{1q}\frac{\partial}{\partial x} + J_{2q}\frac{\partial}{\partial y}$$

$$a = \frac{2}{2+\eta},\quad b = \frac{3\eta}{2+\eta},\quad c = \frac{2-2\eta}{(2+\eta)\mathrm{Fr}^2},\quad \eta = \frac{\rho_f}{\rho_s},\quad \mathrm{Fr}^2 = \frac{U^2}{gL},$$

ρ_s is the density of the particles and $\mathbf{e}_g$ is a unit vector in the direction of the gravity force (as in the previous formulae). Normalising the particle number density by its initial value n_0 (assumed to be constant over the whole region occupied by the particles), we can write:

$$n_{sj}\left|J_j\right| = 1. \tag{17}$$

These equations were solved subject to the following initial conditions:

$$t = 1: \quad \mathbf{r}_{sj} = \mathbf{r}_{0j}, \quad \mathbf{v}_{sj} = 0, \quad J_{pq,j} = 1\ (p \neq q), \quad \vartheta_{pq,j} = 0.$$

The number of particle trajectories was taken equal to the number of blobs: $M = N$.

The authors of [5] showed some examples of the calculations for different values of parameters η and Fr. They illustrated the time evolution of particle number density for the case when $\rho_s \gg \rho_g$ (i.e. $\eta \to 0$) under the assumption that only the effect of the Stokes drag force needs to be taken into account. In this case, the particles, initially uniformly distributed in space, eventually left the high-vorticity region and formed a ring-shaped particle accumulation zone. The ring radius increased with time and sharp concentration gradients developed on the edge of the particle accumulation zone. It was shown that, in the presence of gravity, the particle accumulation region was deformed and removed from the vortex centre. The particle concentration gradient near the boundary of the particle cloud in this case increased with time.

When the density of particles was less than that of the carrier phase, $\rho_s < \rho_g$ (i.e. $\eta > 1$), the flow pattern became rather different from the one described above. In this case, the contribution of the added mass force became predominant. This force led to particle penetration into the high-vorticity region. The centrifugal force, however, resulted in their displacement away from the vortex centre. As a result, the shape of the particle cloud changed non-monotonically with time. The formation of annular zones of particle accumulation in the high-vorticity region was demonstrated. The effect of gravity on these particles was shown to lead to the formation of a particle cloud, with narrow zones of particle inhomogeneity. This cloud was shown to leave the centre of the vortex, significantly increasing in size with time. In this flow, multiple intersections of particle trajectories were observed. To represent the instantaneous patterns of the total particle number density in the regions of multi-valued particle velocities (folds) an additional recalculation procedure was used. This procedure included checking the sign of Jacobian J and the comparison of the values of particle velocities for particle trajectories when these trajectories became very close. When the signs of J or the velocities of the neighbouring trajectories were different, the total particle density was calculated as the sum of different parts of the 'fold'.

3.1.2. Isothermal two-phase plane jet flow Apart from particle flows in a Lamb vortex, the authors of [5] considered an impulse plane two-phase jet, injected into a quiescent carrier phase during Δt. Only the Stokes drag was taken into account. The initial dimensional particle number density n_0 was assumed to be uniform. The particle number density in the mixture was normalised by this value. Three numbers of vortex blobs N were used in the initial calculations: 400, 1000 and 2000. It was shown that the predicted values of velocity for $N = 1000$ and $N = 2000$ were almost indistinguishable. This allowed the authors of [5] to use $N = 1000$ in all their calculations. The values of the maximal velocities were shown to decrease with time, as expected for the post-formation stage of the vortex pair development. The dependence of the predicted velocities on Re (Re = 50, Re = 100, Re = 1000) was shown to be weak at the initial stage of the process; it became more noticeable at a later stage. The carrier phase and vortex pair velocities increased with Re, which is consistent with the vortex pair theory.

Two initial conditions were considered. The first one corresponded to the case when a single-phase jet was injected into a quiescent two-phase mixture with uniform particle concentration, and the second one referred to the injection of a two-phase jet into the pure carrier phase. For the first case, it was shown that particles accumulated near the edges of the vortex pair. For the second case, the deformation of Lagrangian dispersed phase volumes in the flow was shown to be rather complex. In this case, the distribution of the particle concentration could not be correctly calculated on the basis of conventional approaches.

3.2. A combined Lagrangian, viscous-vortex blob and thermal-blob approach

The approach based on a combination of the conventional Lagrangian method for the dispersed phase and viscous-vortex and thermal-blob methods for the carrier phase was suggested and tested in [7]. In this section the main findings reported in the latter paper are briefly summarised.

As in the case considered in the previous section, this approach was applied to modelling of two processes: the time evolution of a two-phase Lamb vortex and the development of an impulse two-phase jet. In contrast to the previous section, the effects of droplet heating and evaporation were taken into account although based on a rather simplistic model. Hence, the analysis of that paper was focused specifically on droplets rather than on particles (which would have included both solid particles and droplets).

It was assumed that the heat flux is directed towards the droplets. Only evaporation was considered and not droplet heating. Given these simplifications, the evaporation rate is controlled by the heat reaching the surface of the droplet:

$$\dot{m} = \frac{q_s}{H}. \tag{18}$$

The corresponding dimensionless equation can be written in the Lagrangian form:

$$\frac{d\sigma^2}{dt} = \delta \left(T - T_s\right) \Psi_h, \tag{19}$$

where

$$\delta = \frac{8\pi}{3} \frac{\sigma_{ref} L \lambda^* \left(T_\infty - T_{ref}\right)}{m_{ref} U H}, \Psi_h = 1 + 0.3\mathrm{Pr}^{1/3}\mathrm{Re}_s^{1/2}, \;\; \mathrm{Pr} = c_p \mu^*/\lambda^*,$$

δ is the droplet evaporation parameter; m_{ref} is the mass of a droplet of radius σ_{ref}, H is the latent heat of evaporation, λ^* is the dimensional gas thermal conductivity.

3.2.1. Non-isothermal Lamb vortex flow As in the previous section, the Lamb vortex was described by Equations (13) and (14). The initialisation of the viscous-vortex blobs was the same as described in the previous section, except that the blobs were placed on concentric circumferences, with the radii increasing with a fixed step $\Delta r = 0.025$ (rather than the step decreasing towards the vortex centre). This led to a series of rings, with the blobs distributed uniformly in each ring. The strength of the blobs was calculated as the difference in the circulations around the outer and inner radii at $t = t_0$ divided by the number of blobs in the ring. Since Solution (13)-(14) has a singularity at $t = 0$, the initial time instant was taken as $t_0 = 1$. Increasing the number of blobs in the calculations, it was found that a satisfactory accuracy could be achieved with 863 viscous-vortex blobs, and a core size $\varepsilon = c_\varepsilon \sqrt{h}$, where h is the width of the corresponding ring, $c_\varepsilon = 0.7$.

For this number of blobs the algorithm gave satisfactory results not only for the velocity field but also for the derivatives of the velocity. The numerical code was verified by comparison of the predicted results with those inferred from Solution (13)-(14). The closeness of these results shows that the numerical code can be expected to be reliable and can be used for the analysis of other flows.

In the next step, the thermal part of the algorithm was verified. This verification was based on the observation that Equation (5) is equivalent to Equation (6), if $\gamma = \mathrm{Pr}$. Hence, an exact solution for the dimensionless temperature field evolution, corresponding to the diffusion of the localised temperature non-uniformity in the Lamb vortex field, can be described as:

$$T = \frac{\mathrm{Re}}{4\pi t} \exp\left(-\mathrm{Re}\frac{\mathbf{r}^2}{4t}\right) \tag{20}$$

Expression (20) was used to verify the calculations of the temperature field distribution. The initial location of thermal blobs was the same as that of the viscous-vortex blobs. For Re = 100, the initial temperature distribution was set as $T = 25/\pi \exp\left(-25r^2\right), t_0 = 1$. The strength of each thermal blob was the temperature at the blob centre, multiplied by the blob area (the area of a ring divided by the number of blobs in the ring). As in the case of the vorticity field, the results for the temperature were shown not to depend on the number of thermal blobs when this number exceeded 863. The latter number of thermal blobs was used in the analysis of [7].

At the next stage, the droplet parameters were calculated. The droplet evaporation was calculated for Re = 100, Pr = 0.8, $\gamma = 1.33$, $\mathrm{Re}_{s0} = 0$, $\delta = -0.1$, where γ is the specific heat ratio. The initial dimensionless temperature distribution was set as $T = \exp\left(-r^2/0.04\right)$, which implies that $0 < T \leq 1$. The fields of the droplet parameters were presented as functions of the distance to the axis of symmetry. The initial conditions for the dispersed phase were the following:

$$t = t_0 : \ \mathbf{r}_s = \mathbf{r}_0, \ \mathbf{v}_s\left(\mathbf{r}_s\right) = \mathbf{v}\left(\mathbf{r}_s\right), T_s\left(x, y\right) = 0, \ \sigma = 1.$$

Inertial droplets were shown to have a tendency to move away from the centre of the vortex. The droplets initially positioned closer to the vortex centre had higher initial velocities and they overtook the droplets initially located at the periphery. The trajectory intersections resulted in the formation of a multivalued field of droplet parameters. The central region of the flow corresponded to the highest difference in the temperatures between the carrier and the dispersed phases, which led to the highest evaporation rates. The effect of the evaporation process on the drag force coefficient was not explicitly taken into account, but this drag force reduced, as droplet radii decreased. As droplets evaporated, their inertia decreased; the droplet velocities were closer to those of the carrier phase in the case of evaporating droplets, compared with the case of non-evaporating droplets.

3.2.2. Non-isothermal plane two-phase jet flow In what follows, the results of application of the same model as described in the previous section to the analysis of the injection of a cold, two-phase jet into a hot, quiescent gas are described. This study was focused on the formation and dynamics of a two-phase vortex pair.

The injection of a water spray with $L = 0.01$ m (nozzle width), $U = 1.5$ m/s, $T^*_\infty - T^*_{ref} = 50$ K (superscript * indicates dimensional temperatures) was considered. A Cartesian coordinate system with the x-axis directed along the jet axis of symmetry, and the origin at the inlet of the jet was used.

In the calculations, the initial dimensionless vorticity distribution corresponded to a smooth step-like velocity profile in the form:

$$u_0 = \frac{1}{1 + \exp\left(-100\left(y + 0.4\right)\right)} + \frac{1}{1 + \exp\left(100\left(y - 0.4\right)\right)} - 1, \ v_0 = 0. \tag{21}$$

The initial dimensionless temperature distribution was taken as $T_0\left(x, y\right) = u_0\left(x, y\right)$. The initial conditions for the dispersed phase were assumed to be the following:

$$t = t_0 : \ \mathbf{r}_s = \mathbf{r}_0, \ \mathbf{v}_s = \left(u_{s0}, v_{s0}\right) = \left(0.8, 0\right), \ T_s\left(x, y\right) = 1, \ \sigma = 1.$$

The vorticity inferred from the velocity distribution described by System (21) was zero in most of the domain, including the axis $y = 0$, but excluding the two small regions at negative and positive y. This allowed the authors of [7] to initially position the viscous blobs in two rectangles $[-4; 0] \times [0.24; 0.56]$ and $[-4; 0] \times [-0.56; -0.24]$. The thermal blobs were located initially in the rectangle $[-4; 0] \times [-0.5; 0.5]$. The domain geometry ($D/L = 4$) corresponded to the vortex pair 'formation number'. The droplets occupied the rectangle $[-4; 0] \times [-0.4; 0.4]$. At the initial time instant, the regions on both sides of the injected liquid column were free of droplets, and the droplet velocities were 20% lower than those of the carrier phase. At the initial instant of time, all blobs and droplets were distributed uniformly. The strength of a blob was calculated as the product of the corresponding value of the parameter (vorticity or temperature) and the blob area. The variable core size for each blob was calculated as $\varepsilon_i = c_\varepsilon \min\limits_{j=1,N}\left(\sqrt{|\mathbf{r}_{vj} - \mathbf{r}_{vi}|}\right)$ for viscous-vortex blobs, and $\varepsilon_i = c_\varepsilon \min\limits_{j=1,M}\left(\sqrt{|\mathbf{r}_{Tj} - \mathbf{r}_{Ti}|}\right)$, for thermal blobs, where $c_\varepsilon = 1.01$.

After the injection, the viscous-vortex blobs concentrated in two spirals, swirled in the opposite directions and propagating into the quiescent gas. The initial locations and velocities of thermal blobs differed from those of viscous-vortex blobs and they fill a region different from the one occupied by the viscous-vortex blobs.

The numerical results for the carrier phase velocity fields were compared with the results predicted by the asymptotic analytical solution to the problem of the diffusion of a vortex pair, according to which the 'centres' of the vorticity regions travel with velocity:

$$V_{VC}\left(\frac{t^*\Gamma_{\text{ini}}}{l_0^2}\right) = \frac{\Gamma_{\text{ini}}}{4\pi l_0}\left[1 - 8.736\left(\frac{t^*\nu}{l_0^2}\right)^2 + 1.845\left(\frac{t^*\nu}{l_0^2}\right)^3\right]. \tag{22}$$

The asymptotic solution (22) is valid for high Reynolds numbers $\text{Re}_a = \Gamma_{\text{ini}}/\nu >> 1$. In this equation, Γ_{ini} is the intensity of each of the vortices at the initial time instant, and $\pm l_0$ are the initial ordinates of the vortices. Since the vortex pair was symmetric, only the top vortex position was calculated using the formula:

$$\Gamma_{\text{top}} = \sum_{i:\, y_{vi}>0} \Gamma_i, \quad \mathbf{r}_{VC} = \frac{1}{\Gamma_{\text{top}}} \sum_{i:\, y_{vi}>0} \mathbf{r}_{vi}\Gamma_i \tag{23}$$

The reference values for the vortex centre position and velocity were taken at $t = 4$. For Solution (22) the initial time instant was taken equal to $t_0 = 2.6$. The calculations were performed for 10000 viscous-vortex blobs, 10000 thermal blobs, and 1681 droplets.

The 'thermal centres' were calculated using the formula

$$\Theta_{\text{top}} = \sum_{i:\, y_{vi}>0} \Theta_i, \quad \mathbf{r}_{TC} = \frac{1}{\Theta_{\text{top}}} \sum_{i:\, y_{vi}>0} \mathbf{r}_{Ti}\Theta_i.$$

Vortex centre dynamics predicted by the numerical simulation were compared with the prediction of the asymptotic solution (see (22)) for the cases of Re = 100 and Re = 1000. It was shown that the asymptotic solution (Re $\to \infty$) predicts slightly higher values than the numerical simulations; there was a better agreement between the results for the case of Re = 1000. In both cases the numerical results agreed satisfactorily with the predictions of the analytical solution at the initial times ($0 < t < 10$); at later times the discrepancy increased.

The cases of droplets with $\beta = 5.5$, $\beta = 0.21$, and $\beta = 0.05$ were considered. In Figs. 1-3, reproduced from [7], clouds of droplets at various time instants are presented. In the top parts of the figures, the highest level of shading corresponds to the largest droplets, and the lowest level to the smallest droplets. In the bottom parts of the figures, the highest level of shading corresponds to the largest phase velocity slips, and the lowest level to the smallest velocity difference. As one can see in these figures, after the injection, a vortex pair forms in the carrier

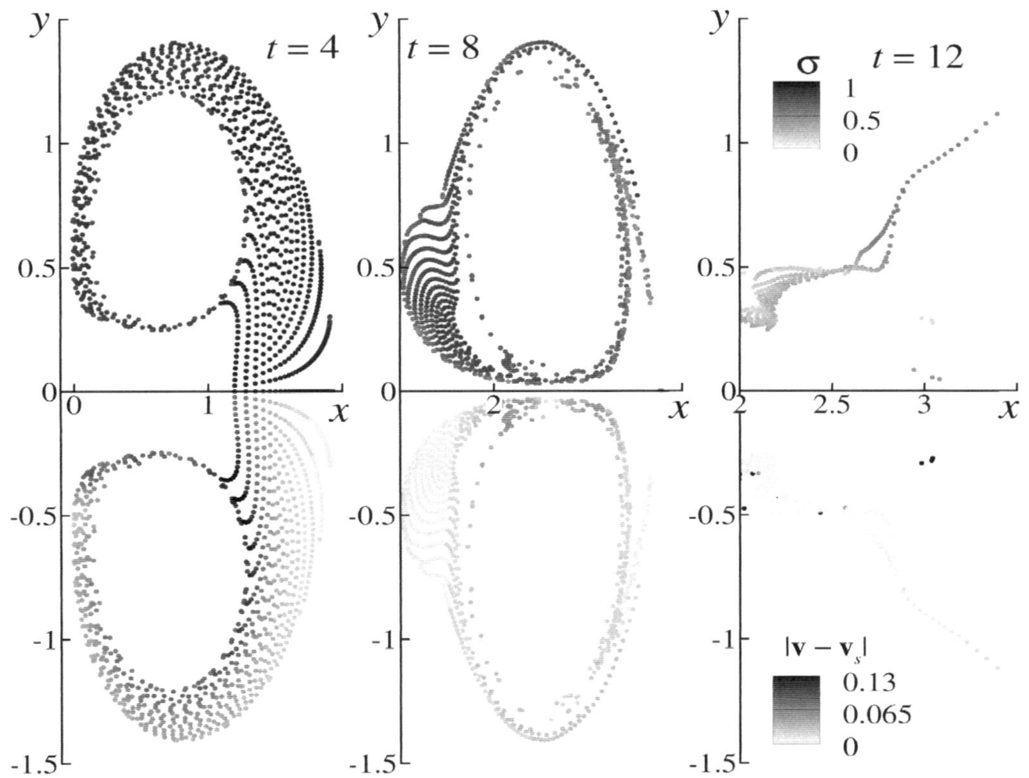

Figure 1. Droplet distributions, their sizes, and velocity slips at $t = 4$, $t = 8$ and $t = 12$; $\beta = 5.5$, $\delta = 0.14$. Reprinted from Ref. [7]. Copyright Elsevier (2016)

phase. The velocities of low-inertia particles ($\beta = 5.5$, see Fig. 1) are quicker to relax than the carrier phase velocities. Thus, the two-phase flow is characterised by better mixing; the droplets form ring-like structures and evaporate more quickly when compared with the cases of larger droplets. At $t = 8$, smaller droplets are located closer to the jet axis, while larger droplets can be seen at the periphery.

The case $\beta = 0.21$ (see Fig. 2) corresponds to the widest two-phase jet among the cases considered. The velocities of the droplets, that were initially close to the inlet, decrease as they enter the quiescent gas; the droplets that enter the flow later can overtake the leading edge. The initially rectangular cloud of droplets is turned inside out after the injection, then it is deformed into a narrow band (see $t = 8$, $t = 12$). The droplets at the sides move more slowly and evaporate more rapidly than those at the centre. As the droplets at the front of the two-phase region evaporate and slow down, larger droplets move to the front edge ($t = 12$). At later times the cloud of droplets is divided into two groups, positioned along the jet axis. The droplet sizes vary within a 7% range.

In the case of droplets with greater inertia ($\beta = 0.05$, see Fig. 3), their sizes remain almost unchanged. The cloud of droplets deforms into a narrow band. The droplets evaporate at a greater rate at the front of the two-phase region. Hence, larger droplets stay at the rear of the droplet cloud. The two-phase region stays closer to the jet axis.

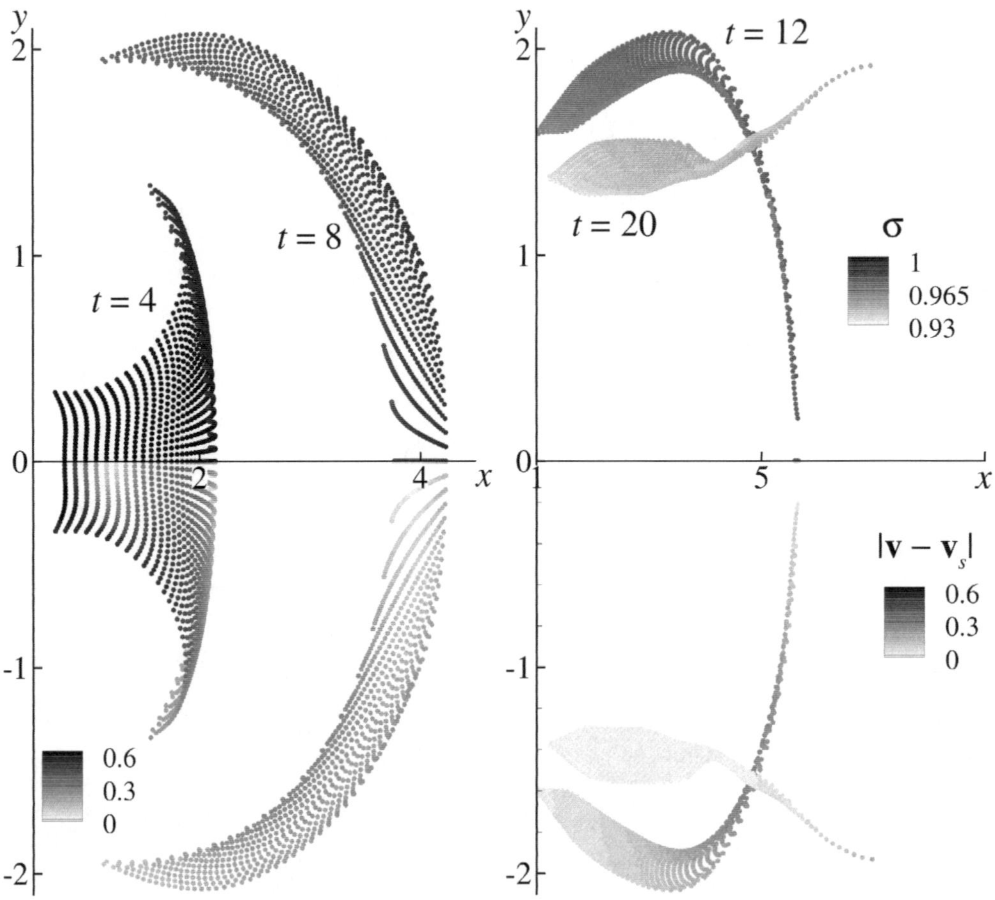

Figure 2. Droplet distributions, droplet diameters, and velocity slips at t = 4, t = 8, t = 12, and t = 20; $\beta = 0.21$, $\delta = 0.006$. Reprinted from Ref. [7]. Copyright Elsevier (2016)

4. Conclusions

New approaches to numerical simulation of two-phase flows suggested in [5, 7] have been discussed and summarised. In [5] an approach to numerical simulation of two-phase flows, based on a combination of the fully Lagrangian method for the dispersed phase and the mesh-free vortex blob method for the carrier phase, was suggested and tested. The one-way coupled, two-fluid approach was used in the analysis. The gas velocity field was restored using the Biot-Savart integral. In this case, the problem of calculation of all parameters in both phases (including particle concentration) is reduced to the solution of a high-order system of ordinary differential equations, describing transient processes in both carrier and dispersed phases. This allows one to study in detail local zones of particle accumulation in complex transient flows, including those with multiple intersections of particle trajectories and the formation of 'folds' in the concentration field of the dispersed phase.

This approach was applied to modelling of two processes: the time evolution of a two-phase Lamb vortex and the development of an impulse two-phase jet. These examples involve the

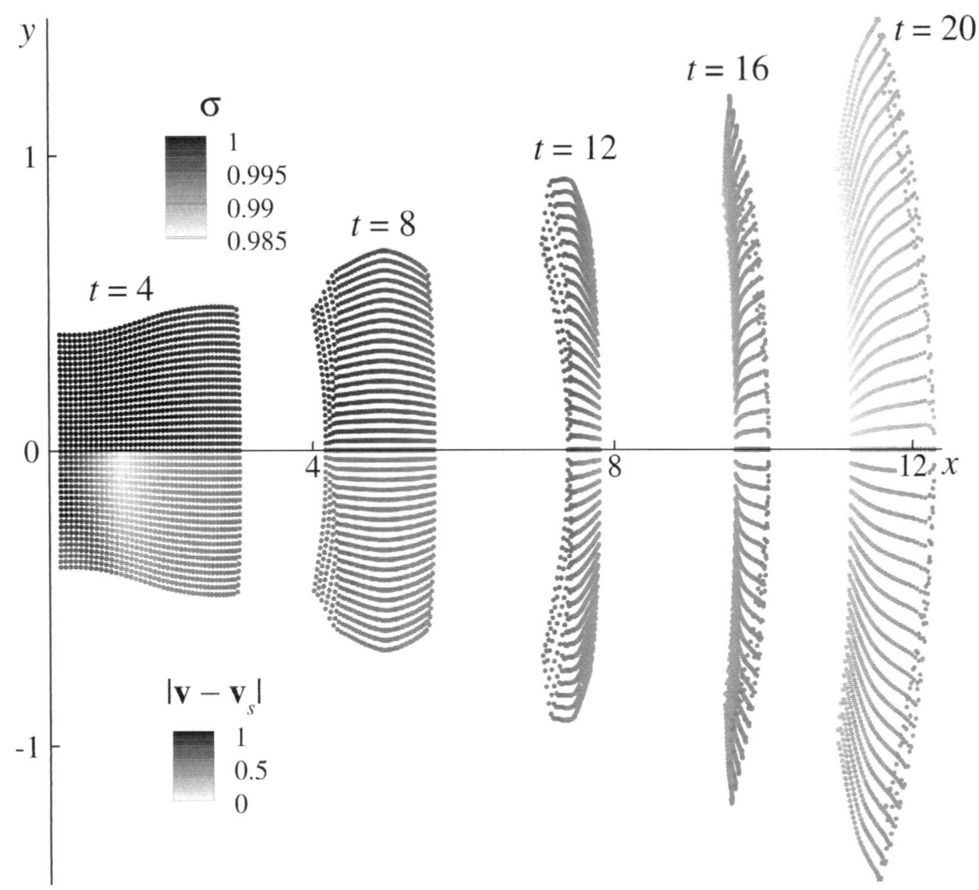

Figure 3. Droplet distributions, droplet diameters, and velocity slips at $t = 4$, $t = 8$, $t = 12$, $t = 16$, and $t = 20$; $\beta = 0.05$, $\delta = 0.0014$. Reprinted from Ref. [7]. Copyright Elsevier (2016)

formation of local zones of particle accumulation and regions of multiple intersection of particle trajectories. These features of the flow cannot be simulated using the conventional Eulerian and Lagrangian methods described in the literature.

In the approach described in [7] the results of the development of a method for modelling of 2D transient, non-isothermal, gas-droplet flows with phase transitions, based on a combination of the viscous-vortex and thermal-blob methods for the carrier phase with the Lagrangian approach for the dispersed phase, were presented. As in [5], the one-way coupled, two-fluid approach was used in the analysis and the gas velocity field was restored using the Biot-Savart integral. The numerical algorithm was verified against the analytical solution for a non-isothermal Lamb vortex and some asymptotic results known in the literature. The method was applied to modelling of an impulse two-phase cold jet injected into a quiescent hot gas, taking into account droplet evaporation. Various flow patterns were obtained in the calculations, depending on the initial droplet size: (i) low-inertia droplets, evaporating at a higher rate, formed ring-like structures and were accumulated only behind the vortex pair; (ii) large droplets moved closer to the jet axis, with their sizes remaining almost unchanged; and (iii) intermediate-size droplets were accumulated in a curved band whose ends trail in the periphery behind the head of the cloud, with larger droplets being collected at the front of the two-phase region.

Acknowledgments

The authors are grateful to EPSRC (grants EP/K005758/1 and EP/M002608/1) for their financial support.

References

[1] Begg S, Kaplanski F, Sazhin S S, Hindle M, and Heikal M 2009 *International Journal of Engine Research* **10** (4) (2009) 195
[2] Healy D and Young J 2005 *Proceedings of the Royal Society of London A: Mathematical, Physical and Engineering Sciences* **461**(2059) 2197
[3] Osiptsov A N 2000 *Astrophys. Space Sci.* **274** 377
[4] Monaghan J 2012 *Annual Review of Fluid Mechanics* **44** 323
[5] Lebedeva N A, Osiptsov A N and Sazhin S S 2013 *Atomization and Sprays* **23** (1) 47
[6] Sazhin S S 2014 *Droplets and Sprays* (London, Springer)
[7] Rybdylova O, Osiptsov A N, Sazhin S S, Begg S and Heikal M 2016 *International J of Heat and Fluid Flow* **58** 93
[8] Goldfarb I, Goldshtein V, Kuzmenko G, and Sazhin S S 1999 *Combustion Theory and Modelling* **3** 769

Positively invariant manifolds: concept and applications

Sergei S. Sazhin[1], **Elena Shchepakina**[2], **and Vladimir Sobolev**[3]

[1] Sir Harry Ricardo Laboratories, Advanced Engineering Centre, School of Computing, Engineering and Mathematics, University of Brighton, Brighton, BN2 4GJ, UK

[2]Department of Technical Cybernetics, Samara National Research University, 34, Moskovskoye shosse, Samara 443086, Russia

[3] Department of Differential Equations and Control Theory, Samara National Research University, 34, Moskovskoye shosse, Samara 443086, Russia

E-mail: S.Sazhin@brighton.ac.uk, shchepakina@smr.ru, hsablem@gmail.com

Abstract. In many applications of the system order reduction models, including those focused on spray ignition and combustion processes, it is assumed that all functions in corresponding differential equations are Lipschitzian. This assumption has not been checked in most cases and the cases when these functions were non-Lipschitzian have sometimes been overlooked. This allows us to question the results of application of the conventional theory of integral manifolds to some such systems. The aim of this paper is to demonstrate that even in the case of singular perturbed systems with non-Lipschitzian nonlinearities the order reduction can be performed, using a new concept of positively invariant manifolds. This is illustrated by several examples including the problem of heating, evaporation, ignition and combustion of Diesel fuel sprays.

1. Introduction

The modelling of many processes, including spray ignition and combustion processes in various engineering applications, has been based on the application of Computational Fluid Dynamics (CFD) codes [1, 2], although the limitations of this approach have been widely discussed in the literature [3]. An alternative approach to modelling these processes has been based on asymptotic methods for their analysis [4], applicable in the case when differences in the rates of change of variables are large. These methods can effectively complement the conventional approach to the problem, based on CFD modelling, by highlighting the physical background of individual processes [4]. One of these methods has been based on the theory of integral manifolds for singularly perturbed systems [5, 6, 7, 8], focused on the analysis of the following autonomous system:

$$\dot{x} = f(x, y, \varepsilon), \quad \varepsilon\dot{y} = g(x, y, \varepsilon) \tag{1}$$

where $0 < \varepsilon \ll 1$, $x \in \mathcal{R}^m, y \in \mathcal{R}^n$, in $\mathcal{R}^{m+n} = \mathcal{R}^m \times \mathcal{R}^n$. A surface $y = \aleph(x, \varepsilon)$ is called a slow invariant manifold of System (1) if any trajectory $x = x(t, \varepsilon)$, $y = y(t, \varepsilon)$ of System (1) that has at least one common point $x = x_0$, $y = y_0$ with the surface $y = \aleph(x, \varepsilon)$, i.e. $y_0 = \aleph(x_0, \varepsilon)$, lies entirely on this surface, i.e. $y(t, \varepsilon) = \aleph(x(t, \varepsilon), \varepsilon)$. In order to find this manifold, the functions $f(x, y, \varepsilon)$ and $g(x, y, \varepsilon)$ must be smooth and therefore satisfy the Lipschitzian condition:

$$\|g(x_1, y_1) - g(x_2, y_2)\| \le \mathcal{L}(\|x_1 - x_2\| + \|y_1 - y_2\|), \tag{2}$$

IOP Conf. Series: Journal of Physics: Conf. Series **811** (2017) 012015 doi:10.1088/1742-6596/811/1/012015

where (x_1, y_1), (x_2, y_2) are arbitrary points in the domain and $\mathcal{L} > 0$.

In numerous papers, including [9, 4], the results of application of this theory to modelling of spray ignition and combustion processes are described. The analyses of spray ignition and combustion processes in these papers are based on the Arrhenius chemical model and these processes are indistinguishable from the point of view of modelling. The authors of [10] drew attention to the fact that in the model described in [4], Condition (2) is not satisfied which cast doubt on the validity of the results presented in [4]. In [10] an alternative approach to the analysis of the problem described in [4], using the new concept of positively invariant manifolds and the Tikhonov theorem, has been adopted (also see [3]).

Similarly to [10], our analysis will focus on the investigation of positively invariant manifolds for non-Lipschitzian systems, describing the processes of spray ignition and combustion. In contrast to [10] we will focus not on the model described in [4], but on a more advanced model, taking into account the volumetric absorption of the thermal radiation in droplets, developed in [11]. The positively invariant manifolds and Lyapunov functions will be used. Some preliminary results of our analysis were reported in [12].

2. Preliminary examples

To clarify the difficulties emerging during the analysis of differential systems with non-Lipschitzian nonlinearities and the differences between smooth and non-smooth functions we consider several simple examples.

Recall that function $f(y)$ satisfies the Lipschitzian condition if for some positive $\mathcal{L}$ Inequality (2) is satisfied for all (x_1, y_1) and (x_2, y_2) from the interval under consideration. In this section the analysis will be focused on one-dimensional functions.

Example 1 *(Smooth case).* A solution to the initial value problem

$$dy/dx = -y^2, \quad y(0) = 1$$

can be presented as

$$y(x) = 1/(1 + x), \tag{3}$$

where $y(x) \to 0$ as $x \to \infty$. It is necessary to use an infinite time interval to obtain this result. The plot of $y(x)$ predicted by (3) is shown in Figure 1.

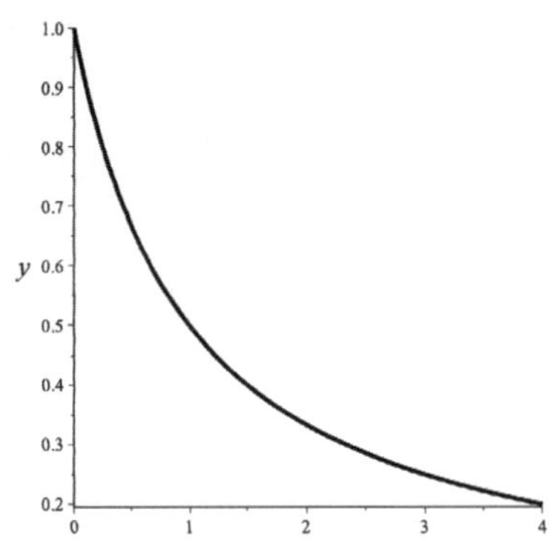

Figure 1. The graph of $y(x)$ predicted by (3) (smooth case).

Example 2 *(Non-Lipschitzian case).* A solution to the initial value problem

$$dy/dx = -y^{1/2}, \quad y(0) = 1$$

can be presented as

$$y(x) = (1 - x/2)^2. \tag{4}$$

This solution is valid at the finite interval $[0, 2]$. This solution is smooth at $x = 2$ but function $F(y) = y^{1/2}$ does not satisfy the Lipschitzian condition at any neighbourhood of $y = 0$ as its derivative is infinitely large at $y = 0$. The plot of $y(x)$ predicted by (4) is shown in Figure 2. Equations leading to non-Lipschitzian solutions are referred to as equations with non-Lipschitzian non-linearities.

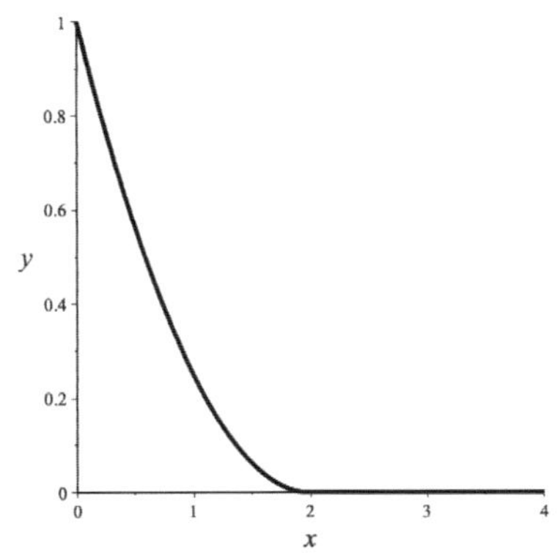

Figure 2. The graph of $y(x)$ predicted by (4) (non-Lipschitzian case).

Example 3 *(Discontinuous case).* A solution to the initial value problem

$$dy/dx = -\text{sgn}(y), \quad y(0) = 1$$

can be presented as

$$y(x) = (1 - x). \tag{5}$$

This solution is valid at the finite interval $[0, 1]$. It is not smooth at $x = 1$. The plot of $y(x)$ predicted by (5) is shown in Figure 3.

These examples show that, in a certain sense, differential equations with non-Lipschitzian non-linearities occupy an intermediate position between equations with smooth and discontinuous non-linearities (i.e. equations with solutions leading to smooth and discontinuous functions).

3. Finite time processes

The physical processes characterised by a finite period of existence are well known (e.g. droplet evaporation time). These processes are commonly described by non-smooth ODEs. In what follows this is illustrated by considering the scalar ODE:

$$dy/dt = f(y), \quad f(0) = 0, \quad y(0) = y_0 > 0.$$

IOP Conf. Series: Journal of Physics: Conf. Series **811** (2017) 012015 doi:10.1088/1742-6596/811/1/012015

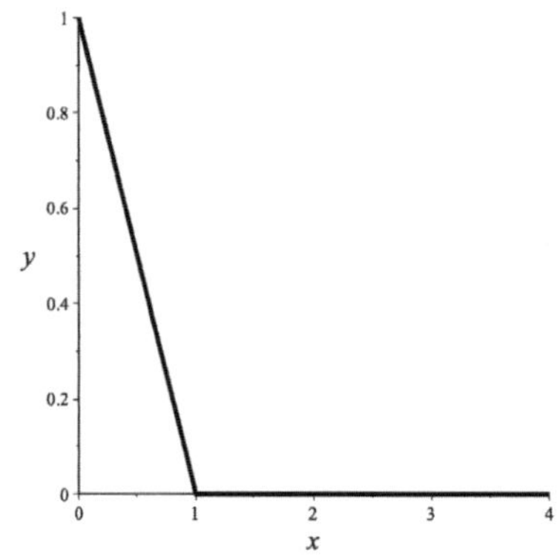

Figure 3. The graph of $y(x)$ predicted by (5) (discontinuous case).

The integration of this equation, assuming that $y(\mathcal{T}) = 0$, where $\mathcal{T}$ is the finite time, gives $\int_{y_0}^{0} \frac{dy}{f(y)} = \mathcal{T}$. Assuming that $f(y) = -y^\alpha$ one can see that the integral in the left hand side of this equation

$$\int_{y_0}^{0} \frac{dy}{f(y)} = \frac{y_0^{1-\alpha}}{1-\alpha}$$

converges if and only if $\alpha < 1$. Thus for a finite $\mathcal{T}$, function $f(y) = -y^\alpha$ does not satisfy the Lipschitzian condition.

This conclusion can be generalised for

$$f(y) = -y^\alpha(\wp + f_1(y)), \tag{6}$$

where $0 < \wp < \infty$, $|f_1(y)| \le \mu < \wp$. This is inferred from the fact that $\wp - \mu \le \wp + f_1(y) \le \wp + \mu$ and, therefore, $\mathcal{T}_- < \mathcal{T} < \mathcal{T}_+$ where

$$\mathcal{T} = \int_{y_0}^{0} \frac{dy}{f(y)}, \quad y(\mathcal{T}) = 0,$$

and

$$\mathcal{T}_+ = \int_{y_0}^{0} \frac{dy}{-y^\alpha(\wp - \mu)} = \frac{y_0^{1-\alpha}}{(1-\alpha)(\wp - \mu)}, \quad \mathcal{T}_- = \int_{y_0}^{0} \frac{dy}{-y^\alpha(\wp + \mu)} = \frac{y_0^{1-\alpha}}{(1-\alpha)(\wp + \mu)}.$$

Example 4 Let us consider the system

$$dx/dt = 1, \quad dy/dt = -y^{1/2}; \quad x(0) = x_0, \quad y(0) = y_0 \tag{7}$$

for which the trajectory $y = 0$ plays the role of a positively invariant manifold, i.e. the trajectory for any solution of (7) with $y_0 = 0$ lies on this manifold for all $t > 0$. The trajectory of any solution of (7) with $y_0 > 0$,

$$y = \left(y_0^{1/2} + (x_0/2) - (x/2)\right)^2,$$

reaches this manifold at $x = 2y_0^{1/2} + x_0$ and then stays on it at longer t. Thus, $y = 0$ is positively invariant but it is not actually invariant since trajectories can leave this manifold when t decreases. An infinite number of trajectories with $y_0 > 0$ can pass through any point on $y = 0$.

IOP Conf. Series: Journal of Physics: Conf. Series **811** (2017) 012015 doi:10.1088/1742-6596/811/1/012015

4. Positively invariant manifolds

The concept of positively invariant manifolds will now be illustrated based on the following system:

$$\dot{x} = f(x, y), \tag{8}$$

$$\dot{y} = \psi(y)g(x, y), \tag{9}$$

where x is a vector and y is a scalar. Scalar function $\psi(y)$ is assumed to be non-Lipschitzian. We assume that $0 < \wp_1 \leq g(x, y) \leq \wp_2$ for sufficiently small non-negative values of y. $f(x, y)$, and $g(x, y)$ are continuous functions. The analysis will focus on $\psi = -y^\alpha$ $(0 < \alpha < 1)$.

As the right hand side of (9) is zero at $y = 0$, any trajectory, described by (8)-(9), starting at $(x_0, 0)$ on the surface $y = 0$ lies on this surface for all $t \geq 0$. This surface is not invariant since not all trajectories of System (8)-(9) which have at least one point in common with this surface lie entirely on it; trajectories can leave this surface when t decreases. This surface, however, is positively invariant since any solution to System (8)-(9) starting at (x_0, y_0) with sufficiently small positive y_0 reaches this surface during a finite time interval. Moreover, this surface is attractive. This can be proven, using the approach suggested in [13], by considering the Lyapunov function $V(y) = y^2/2$ with the derivative $\dot{V}(y) = -y^{1+\alpha}g(x, y)$. $\dot{V}(y) < 0$ for $y > 0$ for all values of x under consideration. This implies the asymptotic stability of $y = 0$ with respect to y ($y \to 0$ as t increases).

The same analysis is applicable to the case when $\psi = (\bar{y} - y)^\alpha$, where $\bar{y}$ is a positive constant.

Example 5 Consider now the initial-value problem

$$\dot{y} = (1 - y)^{1/2}(2 - y)^{1/2}, \tag{10}$$

with $y(0) = y_0$, the solution of which can be written in the form

$$t = \log \frac{2y - 3 + 2\sqrt{(1 - y)(2 - y)}}{2y_0 - 3 + 2\sqrt{(1 - y_0)(2 - y_0)}}. \tag{11}$$

Equation (10) has two equilibria $y = 1$ and $y = 2$. The right hand side of (10), however, has physical meaning only for $y < 1$. Hence, the solution $y(t)$ with $y_0 < 1$ attains the steady state $y = 1$ during the finite time

$$t_1 = \log \frac{1}{3 - 2y_0 - 2\sqrt{(1 - y_0)(2 - y_0)}},$$

which can be found from (11). For example, for $y_0 = 0.5$ this expression gives us

$$t_1 = \log \frac{1}{2 - \sqrt{3}}.$$

A similar situation exists for the system of equations describing spray ignition and combustion which are discussed later.

Example 6 Let us now consider the equation:

$$\dot{y} = C_1(\bar{y} - y)^a(\bar{\bar{y}} - y)^b \exp\left(\frac{y}{1 + \beta y}\right), \quad y(0) = y_0, \tag{12}$$

with the constants $C_1 > 0$, $0 < a < 1$, $0 < b < 1$, $0 < \bar{y} < \bar{\bar{y}}$. This can be considered as a generalisation of Example 5. Recall that the right hand side of Equation (12) has a physical meaning only for $y < \bar{y}$. As in the case of the previous example, solution $y(t)$ with $y_0 < \bar{y}$ attains the steady state $y = \bar{y}$ during a finite time, as shown in Figure 4.

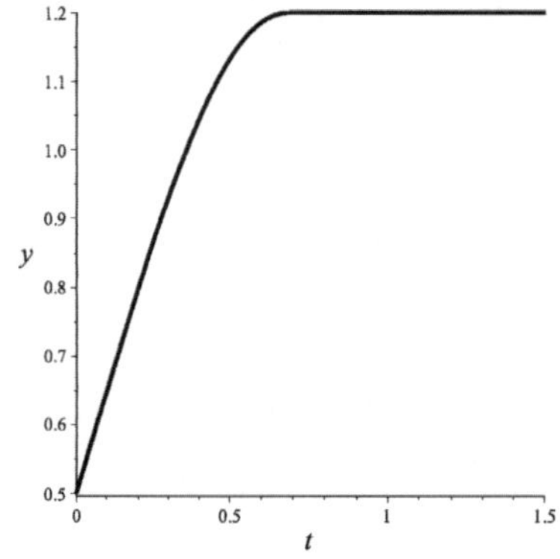

Figure 4. The solution to Equation (12) for $C_1 = 1$, $\beta = 0.1$, $\bar{y} = 1.2$, $\bar{\bar{y}} = 2.1$, $a = 0.5$, $b = 0.25$, $y_0 = 0.5$.

5. Spray ignition and combustion model

Making some natural assumptions, spray heating, evaporation, ignition and combustion processes can be described by the following equations with dimensionless variables and parameters [11]:

$$\frac{d\theta_g}{d\tau} = \frac{1}{\gamma}(P_1(\theta_g, \eta, \xi) - P_2(\theta_g, \theta_d, q^{1/3})), \tag{13}$$

$$\frac{d\eta}{d\tau} = \frac{1}{\tilde{\nu}_f}\left[-P_1(\theta_g, \eta, \xi) + \frac{\psi}{\nu_f}P_{23}(\theta_g, \theta_d, q^{1/3})(1 - \zeta(\theta_d))\right], \tag{14}$$

$$\frac{d\xi}{dt} = -\frac{1}{\tilde{\nu}_{ox}}P_1(\theta_g, \eta, \xi), \tag{15}$$

$$\frac{d\theta_d}{d\tau} = \frac{\varepsilon_2}{\varepsilon_4 q}P_{23}(\theta_g, \theta_d, q^{1/3})\zeta(\theta_d), \tag{16}$$

$$\frac{dq}{d\tau} = -\varepsilon_2 P_{23}(\theta_g, \theta_d, q^{1/3})(1 - \zeta(\theta_d)). \tag{17}$$

where

$$P_1(\theta_g, \eta, \xi) = \eta^a \xi^b \exp\left(\frac{\theta_g}{1 + \beta\theta_g}\right), \quad P_2(\theta_g, \theta_d, r) = \varepsilon_1 r\sqrt{\frac{T_{d0}(1 + \beta\theta_g)}{T_{g0}}}(\theta_g - \theta_d),$$

$$P_3(r) = \frac{\varepsilon_1\varepsilon_3}{4\beta}r^{2+\beta}\left(1 + \beta\theta_g^{\text{ext}}\right)^4, \quad P_{23}(\theta_g, \theta_d, r) = P_2(\theta_g, \theta_d, r) + P_3(r),$$

$$\theta_g^{\text{ext}} = \frac{1}{\beta}\frac{T_{\text{ext}} - T_{d0}}{T_{d0}}, \quad \zeta(\theta_d) = \frac{T_b - T_{d0}\,(1 + \beta\theta_d)}{T_b - T_{d0}},$$

with the initial conditions:

$$\theta_g(0) = \theta_{g0} \neq 0, \quad \theta_d(0) = \theta_{d0} = 0, r(0) = r_0 = 1, \quad \eta(0) = \eta_0, \quad \xi(0) = \xi_0 = 1.$$

Here θ_g, η, ξ, θ_d and r are dimensionless gas temperature, fuel concentration, oxidiser concentration, droplet temperature and droplet radius, respectively; $q = r^3$.

5.1. Reduction of the model

In our previous paper [10], a system of equations similar to the one presented above was considered. The main difference between the physical models considered in [10] and described by System (13)-(17) is that the latter system takes into account the volumetric absorption of thermal radiation in droplets. The analysis of [10] was based on the assumption that the droplet radius is the fastest variable in the model, and on the Tikhonov theorem. The present analysis is based on the concept of positively invariant manifolds and the analysis of the Lyapunov functions.

We can prove that System (13)–(17) has the positively invariant manifold $q \equiv 0$. This proof is based on the observation that this system can be represented as (8), where q plays the role of y and the vector with coordinates θ_g, η, ξ and θ_d plays the role of vector x. The right hand side of (17) can be presented as $-q^{1/3}g(x,q)$ with

$$g(x,q) = \varepsilon_2\left(\varepsilon_1\sqrt{\frac{T_{d0}(1+\beta\theta_g)}{T_{g0}}}(\theta_g-\theta_d) + \frac{\varepsilon_1\varepsilon_3}{4\beta}q^{(1+\beta)/3}\left(1+\beta\theta_g^{\text{ext}}\right)^4\right)(1-\zeta(\theta_d)).$$

The attractivity of this manifold follows from the analysis of Lyapunov function $V(q) = q^2/2$ with derivative $\dot{V}(q) = -q\varepsilon_2 P_{23}(\theta_g, \theta_d, q^{1/3})(1-\zeta(\theta_d))$. $\dot{V}(q) < 0$ for all θ_g, θ_d under consideration. This implies the asymptotic stability of $q = 0$ with respect to q ($q \to 0$ as $t \to +\infty$).

The partial integral of System (13)–(17) is $q = \left(e^{\theta_d}(\zeta(\theta_d))^{\theta_{db}}\right)^{\varepsilon_4}$ (see [11]). Note that $\zeta(\theta_d) \to 0$ as $q \to 0$, i.e. the droplet surface temperature approaches the boiling temperature ($\theta_d \to \theta_{db}$) when $q \to 0$. Having substituted $q = 0$ and $\theta_d = \theta_{db}$ into (13)–(17), the latter can be rewritten as:

$$\frac{d\theta_g}{d\tau} = \frac{1}{\gamma}P_1(\theta_g, \eta, \xi), \tag{18}$$

$$\frac{d\eta}{d\tau} = -\frac{1}{\tilde{\nu}_f}P_1(\theta_g, \eta, \xi), \tag{19}$$

$$\frac{d\xi}{dt} = -\frac{1}{\tilde{\nu}_{ox}}P_1(\theta_g, \eta, \xi). \tag{20}$$

Two integrals

$$\gamma\theta_g + \tilde{\nu}_f\eta = \gamma\theta_{g0} + \tilde{\nu}_f\eta_0 \tag{21}$$

and

$$\gamma\theta_g + \tilde{\nu}_{ox}\xi = \gamma\theta_{g0} + \tilde{\nu}_{ox} \tag{22}$$

of (18)–(20) make it possible to eliminate equations for η and ξ from the following analysis. The final equation for θ_g can be presented as:

$$\frac{d\theta_g}{d\tau} = \frac{1}{\gamma}P_1\left(\theta_g, \eta_0 - \frac{\gamma}{\tilde{\nu}_f}(\theta_g - \theta_{g0}), 1 - \frac{\gamma}{\tilde{\nu}_{ox}}(\theta_g - \theta_{g0})\right). \tag{23}$$

To summarise our analysis, we have shown that application of the concept of positively invariant manifold $q = 0$ for (13)–(17) allows one to perform the order reduction of the original non-Lipschitzian system (13)–(17) and to obtain scalar Equation (23) for θ_g.

6. Conclusions

A concept of positively invariant manifolds for systems with non-Lipschitzian nonlinearities has been introduced and illustrated with several examples. The dynamics of thermal explosion in a fuel droplet/hot air mixture are investigated using this approach. Effects of radiative heating of droplets, semi-transparency of droplets and a limited amount of oxidiser are taken into account. The system of ODEs for gas/droplet temperatures, fuel vapour and oxygen concentrations, and droplet radii are reduced to a single ODE for gas temperature.

7. Acknowledgements

The contribution of V. Sobolev was supported by the Russian Foundation for Basic Research and Samara region(grant 16-41-630529-p) and the Ministry of Education and Science of the Russian Federation as part of a program of increasing the competitiveness of SSAU in the period 2013–2020. E. Shchepakina was supported by the Ministry of Education and Science of the Russian Federation (project RFMEFI58716X0033). S. Sazhin was supported by EPSRC (UK) (grants EP/K020528/1 and EP/M002608/1).

References

[1] Stone, R., "Introduction to Internal Combustion Engines". The MacMillan Press Ltd, Second Edition, 1992.
[2] Sazhina, E.M., Sazhin, S.S., Heikal, M.R., Babushok, V.I. and Johns, R., 2000, *Combustion Science and Technology,* 160, pp. 317-344.
[3] Sazhin, S.S., 2014, "Droplets and Sprays". Springer, London.
[4] Sazhin, S.S., Feng, G., Heikal, M.R., Goldfarb, I., Goldshtein, V. and Kuzmenko, G., 2001, *Combustion and Flame,* 124, 684-701.
[5] Strygin, V.V. and Sobolev, V.A., 1976, *Cosmic Research* 14, pp. 331-335.
[6] Strygin, V.V. and Sobolev, V.A., "Separation of Motions by the Integral Manifolds Method", Nauka, Moscow, 1988 (in Russian).
[7] Gol'dshtein, V. and Sobolev, V.A., 1992, In "Singularity Theory and Some Problems of Functional Analysis", American Mathematical Society, Series 2, pp. 153-173.
[8] Shchepakina, E., Sobolev, V.A. and Mortell, M.P., 2014, "Singular Perturbations: Introduction to System Order Reduction Methods with Applications". Springer Lecture Notes in Math., Vol. 2114.
[9] Goldfarb, I., Gol'dshtein, V., Kuzmenko, G. and Sazhin, S.S., 1999, *Combustion Theory and Modelling,* 3, 769-787.
[10] Sazhin, S.S., Shchepakina, E. and Sobolev, V., 2010, *Mathematical and Computer Modelling,* 52, pp. 529-537.
[11] Goldfarb, I., Goldshtein, V., Katz, D. and Sazhin, S.S., 2007, *International J. of Thermal Sciences,* 46, pp. 358-370.
[12] Sobolev, V.A., Shchepakina, E. and Sazhin, S.S., 2016, Order reduction in non-Lipschitzian models of spray ignition. Proceedings of the 27th European Conference on Liquid Atomization and Spray Systems. Paper P-02, Editors: G. de Sercey and S.S. Sazhin.
[13] Oziraner, A.S. and Rumyantsev, V.V., 1972, *J. Appl. Math. Mech.,* 36, pp. 341-362.

Stable/unstable slow integral manifolds in critical cases

Elena Shchepakina

Department of Technical Cybernetics, Samara National Research University, 34, Moskovskoe shosse, Samara, 443086, Russian Federation.34, Moskovskoye shosse, Samara 443086, Russian Federation

E-mail: shchepakina@smr.ru

Abstract. The paper deals with the problem of a construction of global stable/unstable slow integral manifolds of the singularly perturbed systems in critical cases. In addition to the well-known critical cases a novel scenario of the stability change of the slow integral manifold is considered. All three critical cases leading to the change of the stability are discussed via the Hindmarsh-Rose dynamic model. It is shown that the suitable choice of the additional parameters of the system yields the slow integral manifold with multiple change of its stability.

1. Introduction

The usual assumption of the singular perturbation theory is based on the fact, that the spectrum of the main functional matrix of the fast subsystem lies outside the imaginary axis. However, in many applications this condition is violated that implies the critical phenomena. A violation of this condition can lead to a delaying effect of loss of stability. Phenomenon of the delay of the loss of stability is based on the fact that the actual escape of the phase point from the position of equilibrium, which lost its stability, does not occur immediately. Two scenarios for delaying of the loss of stability are well-known [1, 2].

The first scenario is connected to the critical case, when one real eigenvalue of the linearized fast subsystem passes through zero under a changing of the slow variables. This scenario is associated with the canards or duck-trajectories [3]–[17]. The second scenario one can observe in the critical case, when a pair of complex conjugate eigenvalues passes from the left open complex half-plane to the right one [18, 19, 20].

In present paper the third critical case is considered, when the real parts as well as the imaginary parts of a pair of complex conjugate eigenvalues become zero with the appearance of multiple zero root and then by the birth of a pair of real eigenvalues of opposite signs. All three scenarios of change of stability are discussed via the Hindmarsh-Rose model of nerve conduction [21].

2. Loss of stability of integral manifolds of slow-fast systems

Consider the autonomous singularly perturbed system

$$\frac{dx}{dt} = f(x, y, \mu, \varepsilon), \tag{1}$$

$$\varepsilon\frac{dy}{dt} = g(x, y, \mu, \varepsilon), \tag{2}$$

where x and y are vectors in Euclidean spaces, ε is a small positive parameter, μ is a vector of parameters, vector-functions f and g are sufficiently smooth and their values are comparable to unity. The slow and fast subsystems are described by (1) and (2), respectively.

In many application it is necessary to consider the behaviour of the system as a whole rather than separate trajectories, investigating the system dynamics by methods of the integral manifolds theory (see, for example, [8, 10, 16] and references therein). Especially this approach is effective for analysis of dynamic models with unfixed initial (boundary) conditions and parameters.

Definition 1 *A smooth surface S_ε is called an integral manifold of the system (1), (2) if any trajectory of the system that has at least one point in common with S_ε lies entirely on S_ε.*

Among the integral manifolds we distinguish the invariant surfaces of slow motions whose dimension is equal to that of the slow subsystem, the so-called *slow integral manifolds.* The stability or instability of the slow integral manifold is defined by the stability or instability of its zero-order approximation ($\varepsilon = 0$), the so-called slow surface.

Definition 2 *The surface S described by the equation*

$$g(x, y, \mu, 0) = 0 \tag{3}$$

is called a slow surface. When the dimension of this surface is equal to one, it is called a slow curve.

Let a vector-function $y = \varphi(x, \mu)$ is an isolated root of the equation (3).

Definition 3 *The subset of S is stable (or attractive) if the spectrum of the Jacobian matrix*

$$J = \frac{\partial g}{\partial y}(x, \varphi(x, \mu), \mu, 0) \tag{4}$$

is located in the left open complex half-plane. If there is at least one eigenvalue of the Jacobian matrix (4) with a positive real part then the subset of the slow surface is unstable (or repulsive).

As noted above the slow surface can be considered as a zero-order approximation of the slow integral manifold, hence, in an ε–neighborhood of a stable (unstable) subset of the slow surface there exists a stable (unstable) slow integral manifold.

A slow integral manifold can change its stability in some specific cases. Some critical cases leading to the stability change is considered below. Introduction of the additional conditions allows us to construct of the stable/unstable slow integral manifolds in these critical cases.

2.1. The case of a zero root

Consider the system (1), (2) with scalar parameter μ, for which an equilibrium of the fast equation (2) becomes unstable with transition of one real eigenvalue of (4) through zero when the slow variables are changed. This means that the slow integral manifold of the system loses its stability when the slow variables reach a so-called *breakdown surface* [10, 16].

Definition 4 *The subset of S given by*

$$det\left|\frac{\partial g}{\partial y}(x, y, \mu, 0)\right| = 0 \tag{5}$$

determines a breakdown surface (breakdown, turning or jump points in the scalar case).

The presence of the additional scalar parameter in the differential system provides the possibility of gluing the stable and unstable slow integral manifolds at one point of the breakdown surface to form a single trajectory, the *canard.*

Definition 5 *A canard is a trajectory of a singularly perturbed system of differential equations if it follows at first a stable integral manifold, and then an unstable one. In both cases the length of the trajectory is more than infinitesimally small.*

The technique of a canard construction by a gluing the stable and unstable slow integral manifolds at one point of the breakdown surface was first proposed in [4, 5]. The mathematical justification of this approach for the case when the slow integral manifolds of the system (1), (2) can be represented as $y = h(x, \mu, \varepsilon) = \varphi(x, \mu) + O(\varepsilon)$, $\varepsilon \to 0$, where $h(x, \mu, \varepsilon)$ is a sufficiently smooth function of ε, and x and y are scalar, one can find in [7, 17].

The canards and corresponding values of the parameter μ allow asymptotic expansions in powers of the small parameter ε. Near the slow curve the canards are exponentially close, and have the same asymptotic expansion in powers of ε. An analogous assertion is true for corresponding parameter values. Namely, any two values of the parameter μ for which canards exist have the same asymptotic expansions, and the difference between them is given by $\exp(-1/c\varepsilon)$, where c is some positive number. In this sight we can state the uniqueness for the canard (and the corresponding parameter value) of a planar system.

But in the case $\dim x \geq 2$ the situation is essentially another: if the differential system has a canard then it has one-parameter family of canards at once, and a choice of a value of an additional parameter μ means a selection of a point on the breakdown surface at which the stable and unstable integral manifolds are glued. The mathematical justification of this fact for the case $\dim x \geq 2$, $\dim y = 1$ one can find in [6, 10].

Consider a more general case, when both slow and fast variables are vectors. Let the autonomous system of ordinary differential equations with a small positive parameter ε and an additional scalar parameter μ for the variables x, y, and z is reduced by the standard procedure of elimination of the independent variable t to the nonautonomous form

$$\frac{dy}{dx} = Y(x, y, z_1, z_2, \varepsilon), \; y \in \mathbb{R}^n, \; x \in \mathbb{R}; \tag{6}$$

$$\varepsilon \frac{dz_1}{dx} = 2xz_1 + \mu + Z_1(x, y, z_1, z_2, \varepsilon), \; z_1 \in \mathbb{R}, \; |z_1| \leq r; \tag{7}$$

$$\varepsilon \frac{dz_2}{dx} = A(x)z_2 + \mu B + Z_2(x, y, z_1, z_2, \varepsilon), \; z_2 \in \mathbb{R}^m, \; \|z_2\| \leq r. \tag{8}$$

Here $A(x)$ is a bounded matrix satisfying the Lipschitz condition, its eigenvalues $\lambda_i(x)$ satisfy the condition

$$\mathrm{Re}\, \lambda_i(x) \leq -2\beta < 0 \quad (i = 1, 2, \ldots, m),$$

μ is a constant satisfying

$$|\mu| \leq \varepsilon^2 K, \tag{9}$$

B is a constant vector, and Y, Z_1, and Z_2 are continuous functions satisfying the inequalities

$$\|Y(x, y, z_1, z_2, \varepsilon)\| \leq N, \tag{10}$$

$$|Z_1(x, y, z_1, z_2, \varepsilon)| \leq M\left(\varepsilon^2 + \varepsilon\|z\| + \|z\|^2\right), \tag{11}$$

$$\|Z_2(x, y, z_1, z_2, \varepsilon)\| \leq M\left(\varepsilon^2 + \varepsilon\|z\| + \|z\|^2\right), \tag{12}$$

$$\|Y(x, y, z_1, z_2, \varepsilon) - Y(x, \bar{y}, \bar{z}_1, \bar{z}_2, \varepsilon)\| \leq M(\|y - \bar{y}\| + \|z - \bar{z}\|), \tag{13}$$

$$|Z_1(x,y,z_1,z_2,\varepsilon)-Z_1(x,\bar{y},\bar{z}_1,\bar{z}_2,\varepsilon)|\le M\{(\varepsilon+\|\tilde{z}\|)\|z-\bar{z}\| \\ +(\varepsilon^2+\varepsilon\|\tilde{z}\|+\|\tilde{z}\|^2)\|y-\bar{y}\|\}, \quad (14)$$

$$\|Z_2(x,y,z_1,z_2,\varepsilon)-Z_2(x,\bar{y},\bar{z}_1,\bar{z}_2,\varepsilon)\|\le M\{(\varepsilon+\|\tilde{z}\|)\|z-\bar{z}\| \\ +(\varepsilon^2+\varepsilon\|\tilde{z}\|+\|\tilde{z}\|^2)\|y-\bar{y}\|\}, \quad (15)$$

$$z=\begin{pmatrix} z_1 \\ z_2 \end{pmatrix},\quad \bar{z}=\begin{pmatrix} \bar{z}_1 \\ \bar{z}_2 \end{pmatrix},\quad \|\tilde{z}\|=\max\{\|z\|,\|\bar{z}\|\},$$

M, N, and K are positive constants, $\|\cdot\|$ denotes the ordinary norm in a vector space, and $|\cdot|$ is the absolute value of a scalar.

According to (11), (12) the slow surface of (6)–(8) is described by $z \equiv 0$. A proper choice of μ provides to pick a point on the breakdown surface $x = 0$ for gluing the stable ($x < 0$) and unstable ($x > 0$) slow integral manifolds [2, 8] represented by

$$z = h(x,y,\mu,\varepsilon),\quad h(x,y,\mu,\varepsilon)=\begin{pmatrix} h_1(x,y,\mu,\varepsilon) \\ h_2(x,y,\mu,\varepsilon) \end{pmatrix},\quad h_1\in\mathbb{R},\quad h_2\in\mathbb{R}^m,$$

where

$$|h_1(x,y,\mu,\varepsilon)|\le\varepsilon^{3/2}q,\ |h_1(x,y,\mu,\varepsilon)-h_1(x,\bar{y},\mu,\varepsilon)|\le\varepsilon^{3/2}\delta\|y-\bar{y}\|,$$

$$\|h_2(x,y,\mu,\varepsilon)\|\le\varepsilon^2 q,\ \|h_2(x,y,\mu,\varepsilon)-h_2(x,\bar{y},\mu,\varepsilon)\|\le\varepsilon^2\delta\|y-\bar{y}\|,$$

with some positive constants q and δ.

By fixing a jump point with coordinates $x = 0$, $y = y^*$ on the breakdown surface $x = 0$ of the slow surface, we separate the trajectory

$$y=\phi(x,\mu,\varepsilon),\quad \phi(0,\mu,0)=y^*,$$

on the slow integral manifold $z = h(x, y, \mu, \varepsilon)$, which having reached the jump point along the stable slow integral manifold continues its movement along the unstable one. The following theorem gives sufficient conditions for the system (6)–(8) to have a canard.

Theorem 1 *Let the conditions (9)–(15) be satisfied. Then there are numbers $\varepsilon_0 > 0$ and q, δ such that, for all $\varepsilon \in (0, \varepsilon_0)$, there exist $\mu = \mu^*(\varepsilon)$ and a canard corresponding to $\mu^*(\varepsilon)$, which passes through the point with $x = 0$, $y = y^*$ on the slow integral manifold.*

Thus, a canard can be defined as an one-dimensional slow integral manifold of variable stability, which arise due to special choice of an additional parameter [8, 10].

2.2. The case of a pair of purely imaginary roots

Consider the analytical slow-fast system, obtaining from the system (1), (2) by the time scaling transformation,

$$\frac{dx}{d\tau}=\varepsilon f(x,y,\mu,\varepsilon), \quad (16)$$

$$\frac{dy}{d\tau}=g(x,y,\mu,\varepsilon), \quad (17)$$

for which a singular point of the equation of fast motions becomes unstable with transition of a pair of eigenvalues through the imaginary axis when the slow variables are changed.

As before, the slow integral manifold changes its stability, but unlike the previous case, the stability boundary is not the breakdown surface (curve or point) because the condition (5) is not fulfilled. This means that the trajectory of the system (16), (17) does not jump from the slow integral manifold immediately as in the previous case, it continues to move along the unstable slow surface for a time of order $O(\varepsilon^{-1})$ after crossing the stability border. And this slow path of the trajectory along the unstable part of the slow surface has a distance of order $O(1)$ as $\varepsilon \to 0$. Only then the trajectory can jump from the slow integral manifold and the transition to the fast movement occurs. This delay phenomenon of loss of stability was firstly investigated via an example in [18] and in the general case was considered in [19].

Note that the trajectories described above behave like canards. However there are some difference between these two phenomena. Canards exist in systems with finite smoothness, while the delays considered above occur only in analytic systems. Further, canards are rare and they exist for an exponentially small interval of values of an additional parameter, whereas for a delay phenomenon of loss of stability it is not necessary to select parameters. In the case of delay a function describing the slow integral manifold has a jump discontinuity while in the canard's case it has an infinite discontinuity. If we want to eliminate the trajectory jump from the slow unstable manifold by a constructing of the global stable/unstable slow integral manifold, we need to remove this discontinuity via a selection of a pair of additional functions.

Consider the slow-fast system autonomous system for the variables x, y, and z, which is reduced by the standard procedure of elimination of the independent variable t to the nonautonomous form

$$\frac{dy}{dx} = \varepsilon Y(x, y, z, \varepsilon), \ y \in \mathbb{R}^n, \ x \in \mathbb{R}; \tag{18}$$

$$\frac{dz}{dx} = A(x)z + \mu(y, \varepsilon) + Z(x, y, z, \mu(y, \varepsilon), \varepsilon), \ z \in \mathbb{R}^2, \ \|z\| \le r; \tag{19}$$

where

$$A(x) = \begin{pmatrix} 2x & \nu \\ -\nu & 2x \end{pmatrix},$$

and vector-functions

$$\mu(y, \varepsilon) = \begin{pmatrix} \mu_1(y, \varepsilon) \\ \mu_2(y, \varepsilon) \end{pmatrix}, \ Z(x, y, z, \mu(y, \varepsilon), \varepsilon) = \begin{pmatrix} Z_1(x, y, z, \mu_1(y, \varepsilon), \mu_2(y, \varepsilon), \varepsilon) \\ Z_2(x, y, z, \mu_1(y, \varepsilon), \mu_2(y, \varepsilon), \varepsilon) \end{pmatrix}$$

are continuous and satisfy the following conditions

$$\|Y(x, y, z, \varepsilon)\| \le N, \tag{20}$$

$$\|Z(x, y, z, \mu(y, \varepsilon), \varepsilon)\| \le M\left(\varepsilon + \varepsilon\|z\| + \|z\|^2\right), \tag{21}$$

$$\|Y(x, y, z, \varepsilon) - Y(x, \bar{y}, \bar{z}, \varepsilon)\| \le M\left(\|y - \bar{y}\| + \|z - \bar{z}\|\right). \tag{22}$$

$$\begin{aligned} &\|Z(x, y, z, \mu(y, \varepsilon), \varepsilon) - Z(x, \bar{y}, \bar{z}, \bar{\mu}(\bar{y}, \varepsilon), \varepsilon)\| \le M\left\{(\varepsilon + \|\tilde{z}\|)\|z - \bar{z}\|\right. \\ &\left.+ (\varepsilon + \varepsilon\|\tilde{z}\| + \|\tilde{z}\|^2)\|y - \bar{y}\| + \varepsilon\|\mu - \bar{\mu}\|\right\}, \ \ \|\tilde{z}\| = \max\{\|z\|, \|\bar{z}\|\}, \end{aligned} \tag{23}$$

$$\|\mu(y, \varepsilon)\| \le \varepsilon K, \ \ \|\mu(y, \varepsilon) - \mu(\bar{y}, \varepsilon)\| \le \varepsilon L\|y - \bar{y}\|, \tag{24}$$

where M, N, K, and L are positive constants.

The following theorem holds [22, 23].

Theorem 2 *Let for (18), (19) the conditions (20)–(24) be satisfied. Then there are numbers* $\varepsilon_0 > 0$ *and* q, δ *such that, for all* $\varepsilon \in (0, \varepsilon_0)$, *there exist a function* $\mu = \mu^*(y, \varepsilon)$ *and the stable/unstable slow integral manifold* $z = h(x, y, \varepsilon)$ *corresponding to* $\mu^*(y, \varepsilon)$ *and satisfying the inequalities*

$$\|h(x, y, \varepsilon)\| \le \varepsilon q, \ \ \|h(x, y, \varepsilon) - h(x, \bar{y}, \varepsilon)\| \le \varepsilon\delta\|y - \bar{y}\|.$$

Note that in the case of delayed stability loss considered in [18, 19, 20] the gluing of the stable and unstable parts of the one-dimensional slow integral manifold for the case $\dim y = 0$ requires two additional parameters. But if $\dim y \ge 1$ then a pair of functions is needed to construct the stable/unstable slow integral manifold.

IOP Conf. Series: Journal of Physics: Conf. Series **811** (2017) 012016 doi:10.1088/1742-6596/811/1/012016

2.3. The case of a multiple zero root

Consider a situation when the real parts as well as the imaginary parts of a pair of complex conjugate eigenvalues of the Jacobian matrix (4) become zero simultaneously. In this case a stable focus of the fast equation (2) can transform into a saddle. It should be noted that this bifurcation is among the non-robust. In such situation a canards-like trajectory can arise.

Let us consider this phenomenon via an example, proposed by V.A. Sobolev:

$$\dot{x} = -1, \quad \varepsilon\dot{y} = z, \quad \varepsilon\dot{z} = -x(y+z) + f(x) + \mu, \tag{25}$$

where $f(x) = \alpha_0 x^3 + \alpha_1 x^2 + \alpha_2 x + \alpha_3$, and α_i $(i = 0, \ldots, 3)$ are constant.

The Jacobian matrix (4) of the system (25)

$$J(x) = \begin{pmatrix} 0 & 1 \\ -x & -x \end{pmatrix},$$

has the eigenvalues

$$\lambda_{1,2} = \frac{-x \pm \sqrt{x^2 - 4x}}{2},$$

and for $x \in (0, 4)$ the equilibrium of the fast subsystem of (25) is a stable focus, for $x > 4$ it is a stable node, and for $x < 0$ it is a saddle. Thus, for $x > 0$ the slow integral manifold of (25) is stable, while for $x < 0$ it is unstable.

For $\mu = -\alpha_3 + \varepsilon^2 2\alpha_0$ the system (25) has a canard-like trajectory

$$\begin{cases} y = a_0 x^2 + a_1 x + a_2, \\ z = -\varepsilon(2a_0 x + a_1), \end{cases} \tag{26}$$

where

$$a_0 = \alpha_0, \quad a_1 = \alpha_1 + \varepsilon 2\alpha_0, \quad a_2 = \alpha_2 + \varepsilon\alpha_1 + \varepsilon^2 2\alpha_0.$$

This trajectory, corresponding to the exact solution of the system (25), is the global stable/unstable slow integral manifold. All other trajectories, starting from an initial point in the basin of attraction of the stable part, follows it and continues their movement along the unstable part of the slow integral manifold for a distance of order $O(1)$ as $\varepsilon \to 0$, see Figure 1.

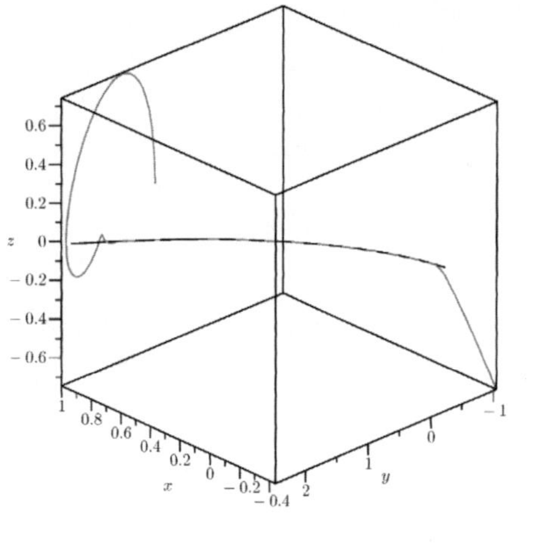

Figure 1. Global stable/unstable slow integral manifold (26) (black line) and the trajectory of the system (25) with initial data $x(0) = 1$, $y(0) = 1$, $z(0) = 0.1$ (red line); $\alpha = 1$, $\beta = 1.2$ $\gamma = 0.1$, $\delta = 1$, $\varepsilon = 0.01$

This result can be generalized.

IOP Conf. Series: Journal of Physics: Conf. Series **811** (2017) 012016 doi:10.1088/1742-6596/811/1/012016

Theorem 3 *For the system (25) with*

$$f(x) = \alpha_0 x^k + \alpha_1 x^{k-1} + \ldots + \alpha_{k-1} x + \alpha_k,$$

where α_i, $(i = 0, \ldots, k)$ and k are constant, there is a number $\varepsilon_0 > 0$ such that, for all $\varepsilon \in (0, \varepsilon_0)$, there exist $\mu = \mu^*(\varepsilon)$ and the exact stable/unstable slow integral manifold

$$\begin{cases} y = a_0 x^{k-1} + a_1 x^{k-2} + \ldots + a_{k-2} x + a_{k-1}, \\ z = -\varepsilon \left(a_0 (k-1) x^{k-2} + a_1 (k-2) x^{k-3} + \ldots + 2a_{k-3} x + a_{k-2} \right), \end{cases}$$

corresponding to $\mu^*(\varepsilon)$, where

$$\begin{aligned} \mu^*(\varepsilon) &= -\alpha_k + \varepsilon^2 2\alpha_{k-3}, \\ a_0 &= \alpha_0, \\ a_1 &= \alpha_1 + \varepsilon(k-1)\alpha_0, \\ a_2 &= \alpha_2 + \varepsilon(k-2)\alpha_1, \\ a_3 &= \alpha_3 + \varepsilon(k-3)\alpha_2 - \varepsilon^2(k-1)(k-2)a_0, \\ a_4 &= \alpha_4 + \varepsilon(k-4)\alpha_3 - \varepsilon^2(k-2)(k-3)a_1, \\ &\ldots \\ a_{k-1} &= \alpha_{k-1} + \varepsilon\alpha_{k-2} - \varepsilon^2 6a_{k-4}. \end{aligned}$$

3. Loss of stability of integral manifolds in the Hindmarsh-Rose model

The considered above scenarios of change of stability of the slow integral manifold occur in the Hindmarsh-Rose model. The model describes the basic properties of individual neurons, the generation of spikes and a constant level potential. In this model, Kirchhoff's law is written for each ionic currents flowing through the cell membrane. The dimensionless form of the Hindmarsh-Rose model is

$$\varepsilon\dot{x} = y - ax^3 + bx^2 - z + I, \tag{27}$$

$$\varepsilon\dot{y} = c - dx^2 - y, \tag{28}$$

$$\dot{z} = s(x - \alpha) - z, \tag{29}$$

where is a transmembrane neuron potential, y and z are the characteristics of ionic currents dynamic, I is ambient current, other parameters reflect the physical features of the neurons. The typical values of the positive parameter are small [24].

The slow curve of system (27)–(29) is described by:

$$\begin{cases} y - ax^3 + bx^2 - z + I = 0, \\ c - dx^2 - y = 0. \end{cases} \tag{30}$$

The graph of the slow curve is shown on Figure 2.

From (30) we get

$$ax^3 + (d - b)x^2 + z - I - c = 0,$$

which allow us to investigate a projection of the slow curve on the XOZ–plane, see Figure 3.

The Jacobian matrix (4) for the system (27)–(29)

$$J = \begin{pmatrix} -3ax^2 + 2bx & 1 \\ -2dx & -1 \end{pmatrix}$$

has a characteristic equation

$$\lambda^2 + \lambda(1 + 3ax^2 - 2bx) + 3ax^2 + 2(d - b)x = 0. \tag{31}$$

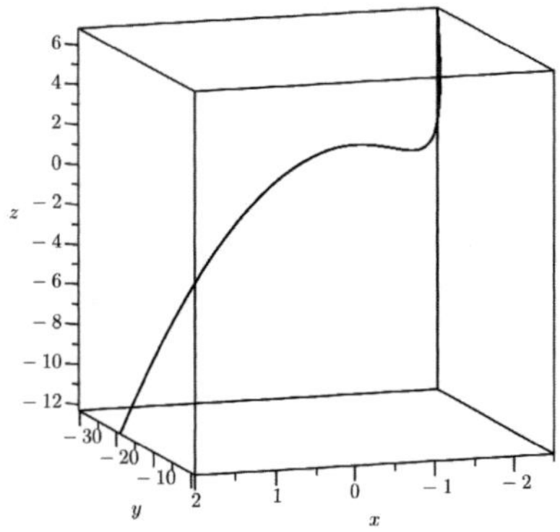

Figure 2. The slow curve (30) of the system (27)–(29); $a = 1$, $b = 3$, $c = 1$, $d = 5$, $I = 2.7$

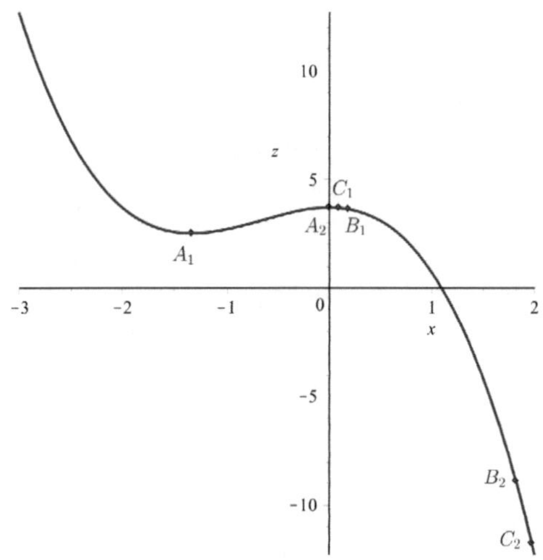

Figure 3. The projection of the slow curve and the points of changing of stability on the XOZ–plane. The parameters' values are the same as for Figure 2

The necessary condition of stability (which is also the sufficient in this case) for the polynomial (31) is

$$\begin{cases} 3ax^2 - 2bx + 1 > 0, \\ 3ax^2 - 2bx + 2dx > 0. \end{cases} \tag{32}$$

From (32) we can determine the abscissas of the points of an expected change of stability of the slow curve:

$$x_1 = \frac{2(b-d)}{3a}, \quad x_2 = 0, \quad x_{3,4} = \frac{b \pm \sqrt{b^2 - 3a}}{3a}.$$

These points divide the slow curve into several parts, see Figure 3.

We check the sign of real part of eigenvalues of the matrix J for all parts to find out whether the region is stable or unstable. One of the two real eigenvalues of the Jacobian matrix of the fast system changes its sign at the points A_1 and A_2, and at these points the slow curve changes its stability. Thus, A_1 and A_2 are the jump points. The proper choice of an additional parameter of the system allows us to glue the stable and the unstable integral manifolds of the system that exist in the ε-neighborhood of the stable and unstable regions of the slow curve. As a result of this gluing we obtain a canard [25]. However, there are other points of the stability's change (see B_1 and B_2 on Figure 3), which are not jump points, because the trajectory does not immediately escape the slow integral manifold as soon as it reaches these points, compared to the previous case. One can find them by equating the real parts of the complex conjugate eigenvalues of J to zero, i.e., from

$$\begin{cases} 3ax^2 - 2bx + 1 = 0, \\ 3ax^2 - 2bx + 2dx > 0. \end{cases}$$

The phenomenon of the delay of loss of stability occurs as the trajectory goes through these points.

In the case under consideration there are points C_1 (between points A_2 and B_1) and C_2 (to the right from the point B_2) on the slow curve, at which the pair of real eigenvalues of the Jacobian matrix J becomes the pair of the complex conjugate eigenvalues.

It is possible to find the relations between parameters values for which the real and the imaginary parts of the pair of complex conjugate eigenvalues become equal to zero simultaneously, i.e.,

$$\begin{cases} 3ax^2 - 2bx + 1 = 0, \\ 3ax^2 - 2bx + 2dx = 0, \end{cases} \tag{33}$$

and a multiple zero root arises with the following emergence of the pair of real eigenvalues with the opposite sign. From (33) we obtain

$$d = \frac{3a}{2\left(b + \sqrt{b^2 - 3a}\right)}, \tag{34}$$

which corresponds to the case when the points A_2, B_2, and C_1 coincide, and the equilibrium of the fast subsystem of (27)–(29) switches from a stable focus to a saddle with decreasing x.

For (34) and the proper choice of the values of other parameters it is possible to construct the global slow integral manifold with multiple change of stability: we need one parameter, say s, to glue the stable and unstable slow integral manifolds at the point A_1 using the techniques of canards (see Figures 4 and 5), and one more, say a or I, for gluing integral manifolds at the point A_2 with help the method described in section 2.3.

As result of these gluing procedures we get the slow integral manifold with multiple change of stability (see Figure 6) that looks like a canard cascade [16, 17, 26]. The difference between these two objects consists in that for a canard cascade we apply the canard technique only.

4. Conclusion

In this paper we outlined the approach for construction the stable/unstable slow integral manifolds of the slow-fast systems of ODE in the critical cases. Novel critical case, in addition to the previously studied cases, leading to the stability change of the slow integral manifold was investigated.

As an illustration of this approach the Hindmarsh-Rose model has been considered, in which all three critical cases of stability change occur. The crucial result of present investigation is that in this dynamic model it is possible to construct the slow integral manifold with multiple change of its stability.

It is important to emphasize that in the present paper we deal with the one-dimensional stable/unstable slow integral manifolds, not necessarily the canards, because these integral manifolds can consist of several trajectories at once (see, for for instance, example 8 in [10]).

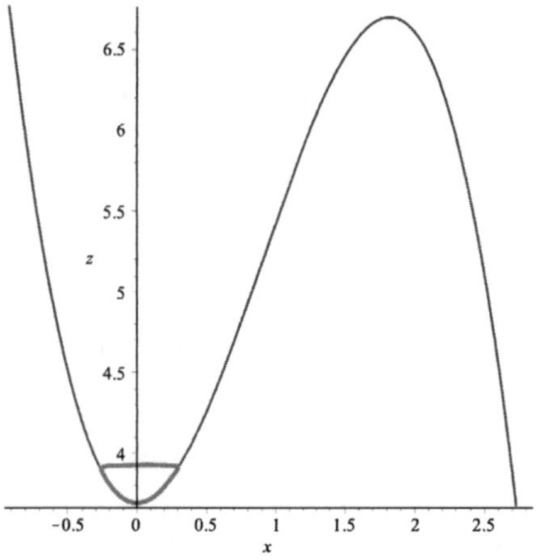

Figure 4. The projections of the canard without head (red line) and the slow curve (blue line) of the system (27)–(29) under the condition (34) and $a = 1$, $b = 3$, $c = 1$, $\alpha = -1.2$, $I = 2.7$, $\varepsilon = 0.01$, $s = 3.0810445478558141214$

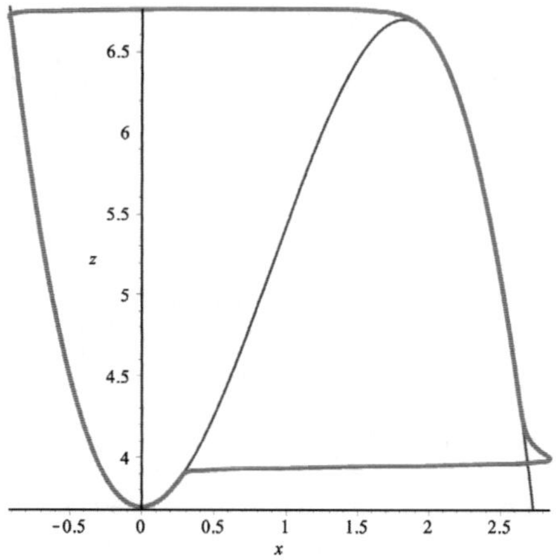

Figure 5. The projections of the canard with head (red line) and the slow curve (blue line) of the system (27)–(29) under the condition (34) and $a = 1$, $b = 3$, $c = 1$, $\alpha = -1.2$, $I = 2.7$, $\varepsilon = 0.01$, $s = 3.0810445478558141213$

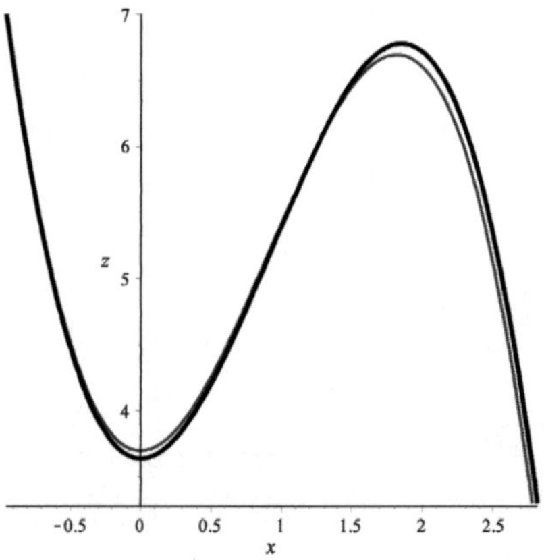

Figure 6. The projections of the integral manifold with multiple change of stability (black line) and the slow curve (blue line) of the system (27)–(29)

Acknowledgment
The author is grateful to Vladimir Sobolev for the proposed example (25). The reported study was funded by RFBR and Samara Region according to the research project 16-41-630529 and the Ministry of Education and Science of the Russian Federation under the Competitiveness Enhancement Program of Samara University (2013-2020).

References

[1] Arnold V I, Afraimovich V S, Il'yashenko Yu S and Shil'nikov L P 1994 *Theory of Bifurcations* (Dynamical Systems vol 5 Encyclopedia of Mathematical Sciences) ed V Arnold (New York: Springer-Verlag)
[2] Shchepakina E A 2004 Two forms of stability change for integral manifolds *Differential Equations* **40**(5) 766–769

[3] Benoit E, Callot J L, Diener F and Diener M 1981–1982 Chasse au canard *Collectanea Mathematica* **31–32** 37–119 [in French]

[4] Gorelov G N and Sobolev V A 1991 Mathematical modelling of critical phenomena in thermal explosion theory *Combustion and Flame* **87** 203–210

[5] Gorelov G N and Sobolev V A 1992 Duck–trajectories in a thermal explosion problem *Applied Mathematics Letters* **5** 3–6

[6] Sobolev V A and Shchepakina E A 1996 *Duck trajectories in a problem of combustion theory* Differential Equations **32** 1177–1186

[7] Gorelov G N, Sobolev V A and Shchepakina E A 1999 *The Singularly Perturbed Models of Combustion* (Samara: SamVien) [in Russian].

[8] Shchepakina E and Sobolev V 2001 Integral manifolds, canards and black swans *Nonlinear Analysis* A **44** 897–908

[9] Shchepakina E A 2003 Black swans and canards in self-ignition problem *Nonlinear Analysis: Real World Appllications* **4**(1) 45–50

[10] Shchepakina E and Sobolev V 2005 *Black Swans and Canards in Laser and Combustion Models* (Singular Perturbation and Hysteresis) ed M P Mortell *et al* (Philadelphia: SIAM) pp 207–255

[11] Shchepakina E 2005 Canards and black swans in model of a 3-D autocatalator *Journal of Physics: Conference Series* **22** 194–207

[12] Shchepakina E and Korotkova O 2011 Condition for canard explosion in a semiconductor optical amplifier *JOSA B: Optical Physics* **28**(8) 1988–1993

[13] Pokrovskii A, Rachinskii D, Sobolev V and Zhezherun A 2011 Topological degree in analysis of cnard-type trajectories in 3-D systems *Applicable Analysis* **90**(7) 1123–1139

[14] Shchepakina E, Korotkova O 2013 Canard explosion in chemical and optical systems *Discrete and Continuous Dynamical Systems B* **18**(2) 495–512

[15] Sobolev V A 2015 Canards and the effect of apparent disappearance *CEUR Workshop Proceedings* **1490** 190–197

[16] Shchepakina E, Sobolev V and Mortell M P 2014 *Singular Perturbations. Introduction to System Order Reduction Methods with Applications* In: Lecture Notes in Mathematics vol. 2114. (Berlin–Heidelber–London: Springer)

[17] Shchepakina E and Sobolev V 2016 Invariant surfaces of variable stability *Journal of Physics: Conference Series* **727** 012016

[18] Shishkova M A 1973 A discussion of a certain system of differential equations with a small parameter multiplying the highest derivatives *Doklady Akademii Nauk SSSR* **209** 576–579 [in Russian]

[19] Neustadt A I 1985 Asymptotic investigation of the loss of stability by an equilibrium as a pair of eigenvalues slowly cross the imaginary axis *Uspekhi Matematicheskikh Nauk* **40** 190–191

[20] Neustadt A I 1987–1988 On delayed stability loss under dynamical bifurcation. I, II *Differential Equations* **23** 2060–2067 **24** 226–233

[21] Hindmarsh J L and Rose R M 1984 A model of neuronal bursting using three coupled first order differential equations *Proceedings of the Royal Society of London B* **221** 87–102

[22] Shchetinina E V 1999 One task of changing the stability of integrated manifolds *Transactions of RANS MMMIC* **3**(3) 129–134 [in Russian]

[23] Schneider K R, Shchetinina E and Sobolev V A 2006 Control of integral manifolds loosing their attractivity in time *Journal of Mathematical Analysis and Applications* **315**(2) 740–757

[24] Hodgkin A L and Huxley A F 1952 A quantitative description of membrane current and its application conduction and excitation in nerve *Journal of Physiology* **117**(4) 500–544

[25] Shchepakina E A 2016 Three scenarios for changing of stability in the dynamic model of nerve conduction *CEUR Workshop Proceedings* **1638** 664–673

[26] Sobolev V 2013 Canard Cascades *Discrete and Continuous Dynamical Systems B* **18**(3) 513–521

One Critical Case in Singularly Perturbed Control Problems

Vladimir Sobolev
Samara National Research University ,
34, Moskovskoye shosse, Samara 443086, Russia
E-mail: hsablem@gmail.com

Abstract. The aim of the paper is to describe the special critical case in the theory of singularly perturbed optimal control problems. We reduce the original singularly perturbed problem to a regularized one such that the existence of slow integral manifolds can be established by means of the standard theory. We illustrate our approach by an example of control problem.

1. Introduction

Consider singularly perturbed differential systems of the type

$$\begin{aligned} \tfrac{dx}{dt} &= f(x, y, t, \varepsilon), \\ \varepsilon \tfrac{dy}{dt} &= g(x, y, t, \varepsilon) \end{aligned} \tag{1}$$

where x and y are vectors, ε is a small positive parameter.

Such systems play an important role as mathematical models of numerous nonlinear phenomena in different fields (see e.g. [1, 2, 3, 4, 5, 6, 7]).

A usual approach in the qualitative study of (1) is to consider first the so called degenerate system

$$\begin{aligned} \frac{dx}{dt} &= f(x, y, t, 0), \\ 0 &= g(x, y, t, 0) \end{aligned}$$

and then to draw conclusions for the qualitative behavior of the full system (1) for sufficiently small ε. In order to recall a basic result of the geometric theory of singularly perturbed systems we introduce the following notation and assumptions for sufficiently small positive ε_0, $0 \le \varepsilon \le \varepsilon_0$.

(A_1). Functions f and g are sufficiently smooth and uniformly bounded together with all their derivatives.

(A_2). There are some region $\mathcal{G} \in R^m$ and a function $h(x, t, \varepsilon)$ of the same smoothness as g such that

$$g(x, h(x, t), t, 0) \equiv 0 \quad \forall (x, t) \in \mathcal{G} \times R.$$

(A_3). The spectrum of the Jacobian matrix $B(x,t) = g_y(x, h(x,t), t, 0)$ is uniformly separated from the imaginary axis for all $(x,t) \in \mathcal{G} \times R$, i.e. the eigenvalues $\lambda_i(x,t)$ $(i = 1, \ldots, n)$ of the matrix $B(x,t)$ satisfy the inequality

$$|Re\lambda_i(x,t)| \geq \gamma \tag{2}$$

for some positive number γ.

Then the following result is valid (see e.g. [8, 9]):

Proposition 1.1. *Under the assumptions* $(A_1)-(A_3)$ *there is a sufficiently small positive* ε_1, $\varepsilon_1 \leq \varepsilon_0$, *such that for* $\varepsilon \in (0, \varepsilon_1)$ *system* (1) *has a smooth integral manifold* $\mathcal{M}_\varepsilon$ *(slow integral manifold) with the representation*

$$\mathcal{M}_\varepsilon := \{(x,y,t) \in R^{n+m+1} : y = \psi(x,t,\varepsilon),\ (x,t) \in \mathcal{G} \times R\}$$

and with the asymptotic expansion

$$\psi(x,t,\varepsilon) = h(x,t) + \varepsilon\psi_1(x,t) + \ldots .$$

The motion on this manifold is described by the *slow* differential equation

$$\dot{x} = f(x, \psi(x,t,\varepsilon), t, \varepsilon). \tag{3}$$

Remark 1.1. *The global boundedness assumption in* (A_1) *with respect to* (x,y) *can be relaxed by modifying* f *and* g *outside some bounded region of* $R^n \times R^m$.

Remark 1.2. *In applications it is usually assumed that the spectrum of the Jacobian matrix* $g_y(x, h(x,t), t, 0)$ *is located in the left half plane. Under this additional hypothesis the manifold* $\mathcal{M}_\varepsilon$ *is exponentially attracting for* $\varepsilon \in I_1$.

The case that assumption (A_3) is violated is called critical. We distinguish three subcases:

(i) The Jacobian matrix $g_y(x,y,t,0)$ is singular on some subspace of $R^m \times R^n \times R$. In that case, system (1) is referred to as a singular singularly perturbed system [10].

(ii) The Jacobian matrix $g_y(x,y,t,0)$ has eigenvalues on the imaginary axis with nonvanishing imaginary parts. A similar case has been investigated in [3, 4, 11].

(iii) The Jacobian matrix $g_y(x,y,t,0)$ is singular on the set $\mathcal{M}_0 := \{(x,y,t) \in R^m \times R^n \times R : y = h(x,t), (x,t) \in \mathcal{G} \times R\}$. In that case, $y = h(x,t)$ is generically an isolated root of $g = 0$ but not a simple one.

Other critical cases were considered, for example, in [3, 4, 12].

The critical case (i) has been treated in [3, 4, 6, 10, 13] and it was considered as applied to the high-gain control problem in [3, 4].

The case (ii) was considered as applied to the manipulator control in [3, 4].

The case (iii) was considered as applied to the partially cheap control problem, see, for example, [3, 4].

It is not inconceivable that combinations of other pairs of critical cases and even thrice critical case are of interest as well.

2. Thrice Critical Case

Consider the control system

$$\varepsilon\dot{x} = A(t,\varepsilon)x + \varepsilon B(t,\varepsilon)u,\ \ x \in R^{n+m},\ \ x(0) = x_0 \tag{4}$$

IOP Conf. Series: Journal of Physics: Conf. Series **811** (2017) 012017 doi:10.1088/1742-6596/811/1/012017

with the cost functional

$$J = \frac{1}{2}x^T(1)Fx(1) + \frac{1}{2}\int\limits_0^1 (x^T(t)Q(t)x(t) + \varepsilon u^T(t)R(t)u(t))dt. \tag{5}$$

where A, F_1, Q are $(n \times n)$-matrices, B is $(n \times m)$-matrix, and R is $(m \times m)$-matrix. Suppose that all these matrices have the following asymptotic presentations with respect to ε:

$A(t,\varepsilon) = \sum\limits_{j\geq 0} \varepsilon^j Aj(t)$, $B(t,\varepsilon) = \sum\limits_{j\geq 0} \varepsilon^j B_j(t)$, $Q(t,\varepsilon) = \sum\limits_{j\geq 0} \varepsilon^j Q_j(t)$, $R(t,\varepsilon) = \sum\limits_{j\geq 0} \varepsilon^j R_j(t)$,
$F(\varepsilon) = \sum\limits_{j\geq 0} \varepsilon^j F_j$,
with smooth on t matrix coefficients, $t \in [0,1]$.

The solution of this problem is the optimal linear feedback control law

$$u = -\varepsilon^{-1}R^{-1}B^TP(t,\varepsilon)x,$$

where P satisfies the differential matrix Riccati equation

$$\varepsilon\dot{P} = -PA - A^TP + PSP - \varepsilon Q, \quad P(1,\varepsilon) = F. \tag{6}$$

Setting $\varepsilon = 0$ we obtain from (6) the matrix algebraic equation

$$-MA_0 - A_0^TM + MS_0M - Q_0 = 0,$$

where $S_0 = B_0R_0^{-1}B_0^T$ and $M = P(t,0)$.

For systems with low energy dissipation the matrices S_0 and Q_0 are equal to zero and the main role plays the linear operator

$$\mathbf{L}X = XA_0 + A_0^TX.$$

For this class of systems the eigenvalues of A_0 are pure imaginary and the spectrum of the linear operator $\mathbf{L}$ has a nontrivial kernel, since sums $(\lambda_i(t) + \lambda_j(t))$, $i,j = 1,\ldots,n$, form its spectrum. This means that the equation (6) is singular singularly perturbed. Thus, the dimension of the slow integral manifold of (6) is greater zero and the problem under consideration is critical in this sense. Moreover, under taking into account that zero eigenvalues are multiple and all other, nonzero eigenvalues of $\mathbf{L}$, are pure imaginary, it is possible to say that this problem is thrice critical.

2.1. *Example*

Let

$$A = \begin{pmatrix} -\varepsilon & 1 \\ -1 & -\varepsilon \end{pmatrix}, B = \begin{pmatrix} 0 \\ 1 \end{pmatrix}, R = (1), Q = \begin{pmatrix} 1 & 0 \\ 0 & 0 \end{pmatrix},$$

$$F = \begin{pmatrix} 0 & 0 \\ 0 & 0 \end{pmatrix}, S = \begin{pmatrix} 0 & 0 \\ 0 & 1 \end{pmatrix}, P = \begin{pmatrix} p_1 & p_2 \\ p_2 & p_3 \end{pmatrix}.$$

Consider the corresponding differential system

$$\begin{aligned} \varepsilon\dot{p}_1 &= 2p_2 + 2\varepsilon p_1 + p_2^2 - \varepsilon, \quad p_1(1) = 0, \\ \varepsilon\dot{p}_2 &= 2\varepsilon p_2 - p_1 + p_3 + p_2p_3, \quad p_2(1) = 0, \\ \varepsilon\dot{p}_3 &= -2p_2 + 2\varepsilon p_3 + p_3^2, \quad p_3(1) = 0. \end{aligned} \tag{7}$$

First, we need to separate it into a slow and a fast subsystem. At first glance, all three equations are singularly perturbed. However, setting $\varepsilon = 0$ we obtain $p_1 = p_2 = p_3 = 0$, and we should consider the matrix of leading terms on the right hand side of the system, which has the form

$$\begin{pmatrix} 0 & 2 & 0 \\ -1 & 0 & 1 \\ 0 & -2 & 0 \end{pmatrix}$$

Obviously, this matrix has a zero eigenvalue and two pure imaginary eigenvalues, i.e. the problem under consideration is twice critical. Moreover, the trivial solution is multiple. This means that we have thrice critical case.

Let $\varepsilon = \mu^2$. Introducing the new variables

$$p_1 = \mu^2 q_1 + \mu, \quad p_2 = \mu^2 q_2 + \mu^2/2, \quad p_3 = \mu^2 q_3 + \mu,$$

and then $s = q_1 + q_3$, we obtain the differential system

$$\begin{aligned} \mu\dot{s} &= 2q_3 + \mu q_2 + 2\mu s + \mu q_2^2 + \mu q_3^2 + 4 + \mu/4, \\ \mu^2\dot{q}_2 &= -s + 2\mu^2 q_2 + 2q_3 + \mu q_2 + \mu^2 q_2 q_3 + \mu/2 + \mu^2, \\ \mu^2\dot{q}_3 &= -2q_2 + 2\mu q_3 + 2\mu^2 q_3 + \mu^2 q_3^2 + 2\mu \end{aligned} \tag{8}$$

with the slow variable s and two fast variables q_2, q_3.

The last system possesses one-dimensional slow invariant manifold which is weakly attractive with respect to argument $1 - t$ because the main matrix of the fast subsystem is

$$\begin{pmatrix} \mu & 2 \\ -2 & 2\mu \end{pmatrix}.$$

Thus, the dimension of the system of Riccati differential equations can be reduced from three to one. Let us construct the slow integral manifold using the fact that it can be asymptotically expanded in powers of the small parameter. Setting

$$q_2 = \varphi(s,\mu) = \mu\varphi_1(s) + \mu^2 \ldots ,$$

$$q_3 = \psi(s,\mu) = \psi_0(s) + \mu\psi_1(s) + \mu^2 \ldots ,$$

we obtain

$$\psi_0(s) = s/2, \quad \varphi_1(s) = s/2, \quad \psi_1(s) = -1/4.$$

Thus we obtain the slow invariant manifold

$$q_2 = \mu s/4 + O(\mu^2), \quad q_3 = s/2 - \mu/4 + O(\mu^2),$$

with the equation on the integral manifold

$$\mu\dot{s} = 4 - \mu/4 + (1 + 2\mu)s + \mu s^2/4 + O(\mu^2).$$

Neglecting by terms of order $O(\mu^2)$ we obtain the scalar solvable equation

$$\mu\dot{s} = 4 - \mu/4 + (1 + 2\mu)s + \mu s^2/4, \quad s(1,\mu) = -2/\mu, \tag{9}$$

and it is possible to obtain the solution of this initial value problem in the analytical form:

$$s = S(t,\mu).$$

Setting for example, $\varepsilon = 0.01$, i.e. $\mu = 0.1$, we obtain

$$S(t) = -24 - \sqrt{417}\tanh(5.105144465t - 5.303589885).$$

For initial variables p_1, p_2, p_3 we obtain the following formulae

$$\begin{aligned} p_1 &= \mu + \mu^2 S(t,\mu) + \mu^3/4 + O(\mu^4), \\ p_2 &= \mu^2/2 + \mu^3 S(t,\mu)/4 + O(\mu^4), \\ p_3 &= \mu + \mu^2 S(t,\mu) - \mu^3/4 + O(\mu^4) \end{aligned} \tag{10}$$

which describe the solution of 10 corresponding to the slow invariant manifold.

The Fig. 1 demonstrates the closeness of solutions of the original system and the system on the slow invariant manifolds for $p_1(t)$. The similar situation takes place for p_2 and p_3, see Fig. 2 and Fig. 3.

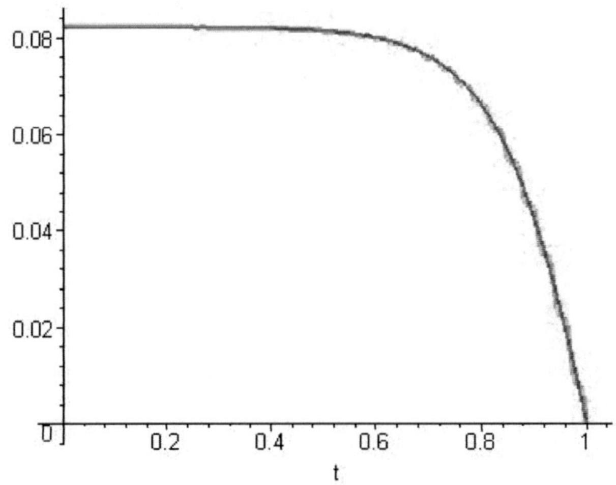

Figure 1. The graphs of p_1 for original differential system (green line) and for the equation on the slow invariant manifold (red line), $\mu = 0.1$.

3. Conclusion

The special critical case for singularly perturbed differential systems are studied in the paper. We have considered singularly perturbed control problem as an application. It has been shown that the reduction of dimensions of these problems can be done by means of the integral manifold method. The slow integral manifolds for the matrix Riccati equation of linear-quadratic control problem are constructed and it is shown that the method of integral manifolds allows us to reduce the dimension of control problems. This approach was used for the investigation of optimal filtering problems in [16, 17].

Acknowledgments

The reported study was funded by RFBR and Samara Region according to the research project 16-41-630524 and the Ministry of education and science of the Russian Federation in the framework of the implementation of Program of increasing the competitiveness of SSAU for 2013–2020 years.

IOP Conf. Series: Journal of Physics: Conf. Series **811** (2017) 012017 doi:10.1088/1742-6596/811/1/012017

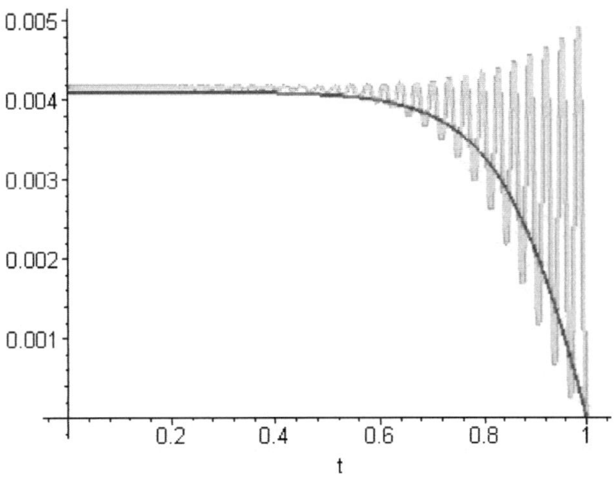

Figure 2. The graphs of p_2 for original differential system (green line) and for the equation on the slow invariant manifold (red line), $\mu = 0.1$.

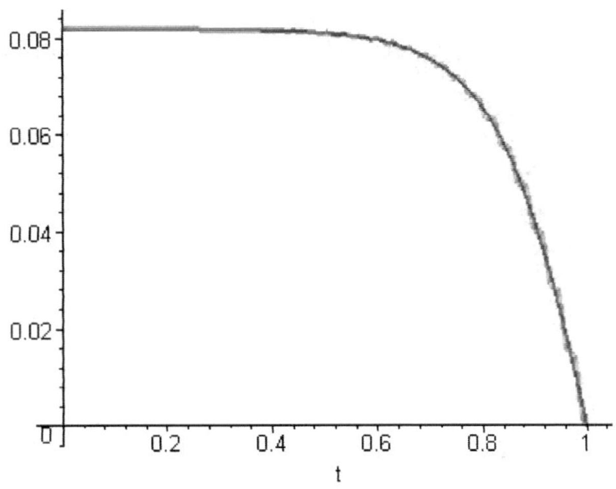

Figure 3. The graphs of p_3 for original differential system (green line) and for the equation on the slow invariant manifold (red line), $\mu = 0.1$.

References

[1] Vasilieva A B, Butuzov V F and Kalachev L V 1995 *The Boundary Function Method for Singular Perturbation Problems* (Philadelphia: SIAM)
[2] O'Malley R E 1991 *Singular Perturbation Methods for Ordinary Differential Equations* Appl. Math. Sci. **89** (New-York: Springer–Verlag)
[3] Mortell M P, O'Malley R E, Pokrovskii A and Sobolev V A 2005 *Singular Perturbation and Hysteresis* (Philadelphia: SIAM)
[4] Shchepakina E, Sobolev V and Mortell M P 2014 *Singular Perturbations. Introduction to System Order Reduction Methods with Applications* (Lect Notes in Math **2114**) (Berlin–Heidelber–London: Springer)
[5] Kokotović P V, Khalil K H and O'Reilly J 1986 *Singular Perturbation Methods in Control: Analysis and Design* (Philadelphia: SIAM)
[6] Vasilieva A B and Butuzov V F 1980 *Singularly Perturbed Equations in the Critical Case* Tech. Report MRC–TSR 2039 (Madison: University of Visconsin)
[7] Sobolev V A 1991 Singular perturbations in linearly quadratic optimal control problems *Autom Remote Control* 52:2 180–189
[8] Strygin V V and Sobolev V A 1976 Effect of geometric and kinetic parameters and energy dissipation on orientation stability of satellites with double spin *Cosmic Research* **14** 331–335
[9] Sobolev V A 1984 Integral manifolds and decomposition of singularly perturbed systems *System and Control Letters* **5** 169–179
[10] Gu Z, Nefedov N N and O'Malley R E 1989 On singular singularly perturbed initial values problems *SIAM J Appl Math* **49** 1–25
[11] Sobolev V A and Strygin V V 1978 Permissibility of changing over to precession equations of gyroscopic systems *Mechanics of Solids* **5** 7–13
[12] Shchepakina, E 2003 Black swans and canards in self–ignition problem. *Nonlinear Anal.: Real World Appl.* **4**, 45–50
[13] Kononenko L I and Sobolev V A 1994 Asymptotic expansion of slow integral manifolds *Siberian Math J* **35** 1119–1132
[14] O'Malley R E, Jameson A 1975-77 Singular perturbations and singular arcs I, II. *Trans Autom Control* AC-20 218–226 and AC-22 328–337
[15] Smetannikova E , Sobolev V 2005 Regularization of Cheap Periodic Control Problems. *Autom Remote Control* 66:6, 903–916 (2005)
[16] Osintsev M S and Sobolev V A 2013 Reduction of dimensionality of optimal estimation and control problems for systems of low dissipativity solid bodies *Autom Remote Control* 74:8 1334–1347
[17] Osintsev M S and Sobolev V A 2014 Reduction of dimension of optimal estimation problems for dynamical systems with singular perturbations *Computational Mathematics and Mathematical Physics* 54:1 45-58

IOP Conf. Series: Journal of Physics: Conf. Series **811** (2017) 012018 doi:10.1088/1742-6596/811/1/012018

An asymptotic solution to a passive biped walker model

Sergey A Yudaev[1,*], **Dmitrii Rachinskii**[2] **and Vladimir A Sobolev**[1]

[1] Samara National Research University, Russia

[2] Department of Mathematical Sciences, The University of Texas at Dallas, USA

*Corresponding author: s.a.yudaev@gmail.com

Abstract. We consider a simple model of a passive dynamic biped robot walker with point feet and legs without knee. The model is a switched system, which includes an inverted double pendulum. Robot's gait and its stability depend on parameters such as the slope of the ramp, the length of robot's legs, and the mass distribution along the legs. We present an asymptotic solution of the model. The first correction to the zero order approximation is shown to agree with the numerical solution for a limited parameter range.

1. Introduction

Passive dynamic walkers have been widely developed since 1990 when they were introduced [1]. The problem of legged robot locomotion continues to generate interest of researchers attempting to improve the design of walkers.

Most of us do walking as a daily routine without any thought or care. Nevertheless, walking is a complex process depending on various factors. Human body, controlled by nervous system, involves skeletal muscles and limbs to reproduce an efficient and natural gait. From there, biped gait dynamics are examined by several disciplines. Passive walkers constitute a class of robots that use gravity to power their dynamics. Several experimental studies have shown that this kind of walking is possible with reasonable stability over a range of slopes without any actuation [2–4]. Further, passive dynamics designs were adopted for models with different kind of actuation strategies [5, 6]. However, relatively few analytic and asymptotic results are available for non-actuated walking. In this paper, we consider a simple biped model, which extends the model that has been comprehensively analyzed in [7]. We follow a similar approach, but extend the study by including additional point masses of the legs and allowing the positions of these masses to vary.

The paper organized as follows. Section 2 contains a brief description and assumptions of the model. Equations of motion and equations of transition are presented in Section 3. Section 4 presents asymptotic results, which are complemented by, and compared to, numerical simulations.

2. Model Description

Figure 1 presents a simple passive walker in two dimensions. Given appropriate initial conditions and placed at the top of the ramp with a small slope, this robot starts moving.

The model consists of two rigid legs. The *sway* leg and the *stance* leg make angles θ_{sw}, θ_{st} with the upward direction, respectively. We consider kneeless legs of mass m and length l connected

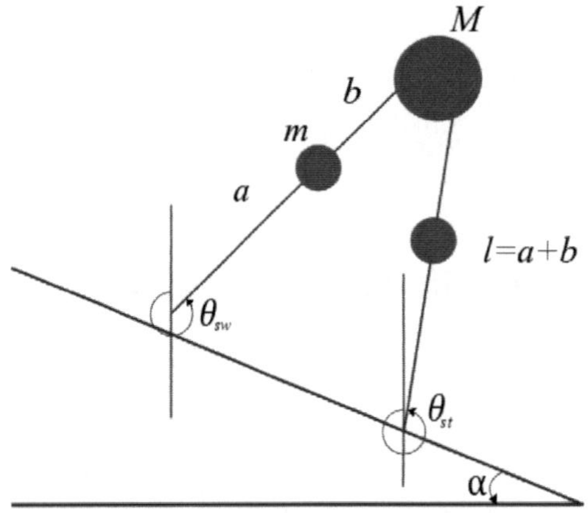

Figure 1. A simple biped robot walker on a slope.

through a hip joint of mass M. The point mass m is placed at a distance a from the hip joint and a distance $b = l - a$ from the foot (tip of the leg). The ramp slope is denoted by α. All the angles are measured counterclockwise.

A few simplifying assumptions make the system better tractable. No friction is included at the leg joint. A perfectly rigid ground is assumed. Foot collision is assumed to be absolutely plastic. As a consequence of this, the foot impact and foot transition are instantaneous. At the time of the impact, t_{imp}, the distance between the foot of the swing leg and the walking surface becomes zero, hence the constraint $\theta_{sw}(t_{imp}) + \theta_{st}(t_{imp}) = -2\alpha \pmod{2\pi}$ is met. At this instant, the feet are renamed (switched), the stance leg becomes the swing leg and vice versa. The stance foot is assumed to be fixed on the ground until the impact occurs. The period of motion between the foot transitions is called the gait cycle.

3. Equations of Motion

The model is hybrid since continuous motion is interrupted by discrete events. More precisely, the continuous motion during the gait cycle is followed by a discrete legs' switch. Equations of motion for this hybrid model were obtained in [8]. Introducing new parameters $\beta = m/M$, $\nu = b/l$, and $\mu = a/l = 1 - \nu$, one can describe the gait cycle as the second order system of the form

$$M(\theta)\ddot{\theta} + N(\theta, \dot{\theta})\dot{\theta} + G(\theta) = 0\,, \tag{1}$$

where the zero right hand side corresponds to passive dynamics. Here $\theta = (\theta_{sw}, \theta_{st})^T$ is the vector of angular coordinates; $M(\theta)$ represents the inertia matrix; the matrix $N(\theta, \dot{\theta})$ accounts for centrifugal and Coriolis forces; and, $G(\theta)$ contains the gravity terms:

$$M(\theta) = \begin{bmatrix} \beta\nu^2 & -\beta\nu\cos\varphi \\ -\beta\nu\cos\varphi & 1+\beta+\beta(1-\nu)^2 \end{bmatrix},$$

$$N(\theta, \dot{\theta}) = \begin{bmatrix} 0 & \beta\nu\dot{\theta}_{st}\sin\varphi \\ -\beta\nu\dot{\theta}_{sw}\cos\varphi & 0 \end{bmatrix},$$

$$G(\theta) = \frac{g}{l}\begin{bmatrix} \beta\nu\sin\theta_{sw} \\ -(1+2\beta-\beta\nu)\sin\theta_{st} \end{bmatrix},$$

where we denote $\varphi = \theta_{st} - \theta_{sw}$.

Since centrifugal and Coriolis forces affects the walker motion only slightly, we ignore the term $N(\theta, \dot{\theta})\dot{\theta}$. Then, equation (1) becomes

$$M(\theta)\ddot{\theta} + G(\theta) = 0\,. \qquad (2)$$

To describe the switch of stance and swing legs, we use the algebraic transition equation

$$\theta^+ = J\theta^- \qquad \text{with} \qquad J = \begin{bmatrix} 0 & 1 \\ 1 & 0 \end{bmatrix},$$

which relates the pre-impact and post-impact coordinate values. Here, the "–" and "+" superscripts refer to the values of state variables before and after the impact, respectively. The conservation of the angular momentum imposes a relation between the pre- and post-impact angular velocities:

$$\dot{\theta}^+ = [Q(\varphi)]^{-1} P(\varphi)\,\dot{\theta}^-,$$

where

$$P(\varphi) = \begin{bmatrix} (1 + 2\beta - 2\beta\nu)\cos\varphi & -\beta(1-\nu)\nu \\ -\beta(1-\nu)\nu & 0 \end{bmatrix},$$

$$Q(\varphi) = \begin{bmatrix} \beta\nu(1-\cos\varphi) & \beta + \beta(1-\nu)^2 + 1 - \beta\nu\cos\varphi \\ \beta\nu^2 & -\beta\nu\cos\varphi \end{bmatrix}.$$

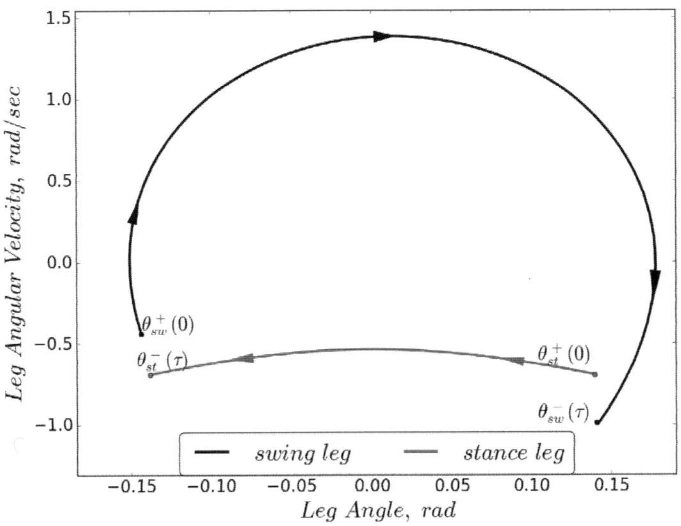

Figure 2. Periodic solution on the phase space diagram.

We reproduced some numerical results from [9] for the hybrid model. Figure 2 presents the phase diagram of one gait cycle for a periodic solution (of period τ). Because the motion is periodic,

$$\theta^+_{st}(\tau) = \theta^+_{sw}(0), \quad \theta^+_{sw}(\tau) = \theta^+_{st}(0), \quad \dot{\theta}^+_{st}(\tau) = \theta^+_{sw}(0), \quad \dot{\theta}^+_{sw}(\tau) = \theta^+_{st}(0).$$

The switch corresponds to the instantaneous transition from $\theta^-_{st}(\tau)$ to $\theta^+_{sw}(0)$ and from $\theta^-_{sw}(\tau)$ to $\theta^+_{st}(0)$.

The time trace of a periodic solution for several gait cycles is presented in Figure 3. Vertical jumps correspond to the switching moments defined by the relationship $\theta_{sw}(t_{imp}) + \theta_{st}(t_{imp}) = -2\alpha \pmod{2\pi}$. Parameters for both Figures 2 and 3 are $\beta = 0.01$, $\nu = 0.2$, $\mu = 0.8$, and $\alpha = 0.0061\ rad$.

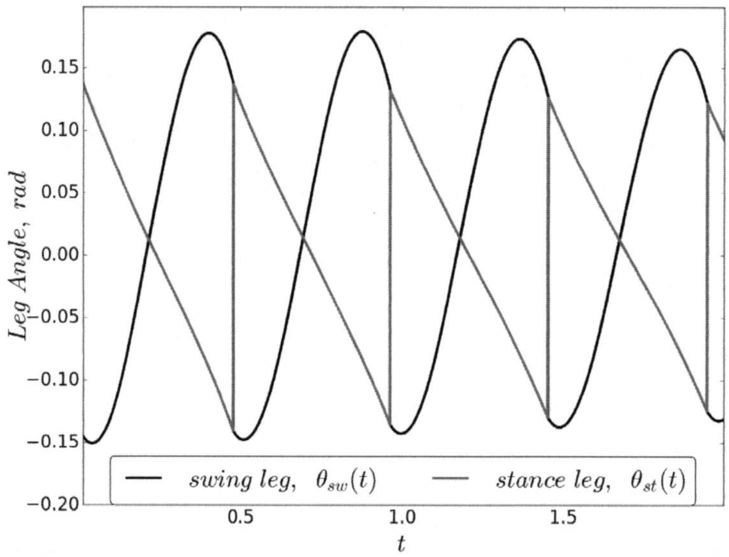

Figure 3. Time trace of the periodic solution.

The dependence of the gait cycle on parameters has been studied in [10].

4. Asymptotic Solution

We use perturbation method in order to study walking cycles for $\alpha \ll 1$ and $\beta \ll 1$. System (2) written componentwise has the form

$$\nu\ddot{\theta}_{sw} - \cos\varphi\ddot{\theta}_{st} + \frac{g}{l}\sin\theta_{sw} = 0\,,$$

$$-\beta\nu\cos\varphi\ddot{\theta}_{sw} + (1 + \beta + \beta(1-\nu^2))\ddot{\theta}_{st} - \frac{g}{l}(1 + 2\beta - \beta\nu)\sin\theta_{st} = 0\,.$$

Rearranging these equations, one obtains

$$q(\varphi)\ddot{\theta}_{st} - \beta\cos\varphi\sin\theta_{sw} - p\sin\theta_{st} = 0\,, \tag{3}$$

$$\nu q(\varphi)\ddot{\theta}_{sw} - p\cos\varphi\sin\theta_{st} - (p(\nu-1) + \nu + \beta\nu)\sin\theta_{sw} = 0\,, \tag{4}$$

where $p = \beta\nu - 2\beta - 1$ and

$$q(\varphi) = (\beta\cos\varphi^2 - 2\beta + 2\beta\nu - \beta\nu^2 - 1)\frac{l}{g}\,.$$

Following [7], we introduce the scaling

$$\alpha = \varepsilon^3, \qquad \theta_{st} = \varepsilon\Theta_{st}, \qquad \theta_{sw} = \varepsilon\Theta_{st},$$

where $\Theta_{sw} \equiv \Theta_{sw}(t)$, $\Theta_{st} \equiv \Theta_{st}(t)$. Substituting these formulas into equations (3), (4) and expanding the equations in a power series with respect the small parameter ε results in two governing equations with no zero order terms. After dividing both equations by ε, the expansion contains only even powers of ε. Therefore, the paremeter $\delta = \varepsilon^2$ is introduced to obtain

$$k\ddot{\Theta}_{st} + p\Theta_{st} + \beta\Theta_{sw} + \left(s(\Theta_{st},\Theta_{sw})\ddot{\Theta}_{st} - \frac{p}{6}\Theta_{st}^3 - \frac{\beta}{2}\Theta_{sw}\Theta_{st}^2 + \beta\Theta_{sw}^2\Theta_{st} - \frac{2\beta}{3}\Theta_{sw}^3\right)\delta = 0\,,$$

$$k\nu\ddot{\Theta}_{sw} + p\Theta_{st} + j\Theta_{sw} + \left(s(\Theta_{st},\Theta_{sw})\nu\ddot{\Theta}_{sw} - p\left(\frac{2}{3}\Theta_{st}^3 - \Theta_{sw}\Theta_{st}^2 + \frac{1}{2}\Theta_{sw}^2\Theta_{st}\right) - \frac{j}{6}\Theta_{sw0}^3\right)\delta = 0\,,$$

where $j = \beta\nu^2 - 2\beta\nu + 2\beta + 1$, $k = (\beta\nu^2 - 2\beta\nu + \beta + 1)l/g$, and

$$s(\Theta_{st},\Theta_{sw}) = \frac{l}{g}\beta(\Theta_{sw} - \Theta_{st})^2\,.$$

Next, we use the regular expansion with respect to two independent small parameters δ and β up to the linear terms:

$$\Theta_{sw} = \Theta_{sw0} + \delta\Theta_{sw1} + \beta\Theta_{sw2}\,, \qquad \Theta_{st} = \Theta_{st0} + \delta\Theta_{st1} + \beta\Theta_{st2}\,. \tag{5}$$

The regular expansion results in the following equations for the zero order approximations:

$$\ddot{\Theta}_{st0} - \frac{g}{l}\Theta_{st0} = 0\,, \tag{6}$$

$$\ddot{\Theta}_{sw0} - \frac{g}{\nu l}(\Theta_{st0} - \Theta_{sw0}) = 0\,; \tag{7}$$

equations for the first corrections with respect to the parameter δ:

$$\ddot{\Theta}_{st1} - \frac{g}{l}\Theta_{st1} = -\frac{g}{6l}\Theta_{st0}^3\,, \tag{8}$$

$$\ddot{\Theta}_{sw1} + \frac{g}{\nu l}\Theta_{sw1} = \frac{g}{\nu l}\left(\Theta_{st1} - \frac{2}{3}\Theta_{st0}^3 + \Theta_{st0}^2\Theta_{sw0} - \frac{1}{2}\Theta_{st0}\Theta_{sw0}^2 + \frac{1}{6}\Theta_{sw0}^3\right); \tag{9}$$

and, equations for the first corrections with respect to the parameter β:

$$\ddot{\Theta}_{st2} - \frac{g}{l}\Theta_{st2} = (1 + \nu - \nu^2)\frac{g}{l}\Theta_{st0} - \frac{g}{l}\Theta_{sw0}\,, \tag{10}$$

$$\ddot{\Theta}_{sw2} + \frac{g}{\nu l}\Theta_{sw2} = (1 - \nu + 1/\nu)\frac{g}{l}\Theta_{st0}(t) + \frac{g}{\nu l}(\Theta_{st2} - \Theta_{sw0})\,. \tag{11}$$

Equations (6)-(11) can be easily solved consecutively.

The functions $\theta_{st}(t) = \varepsilon\Theta_{st}(t)$, $\theta_{sw}(t) = \varepsilon\Theta_{sw}(t)$ obtained from formulas (5)-(11) can be used to approximate the motion during the gait cycle between the moments when the legs are switched. Figure 4 compares this approximation with a numerical solution of the equations of motion (2) for $\delta = \alpha^{2/3} = \varepsilon^2 = 0.0086$, $\beta = 0.01$.

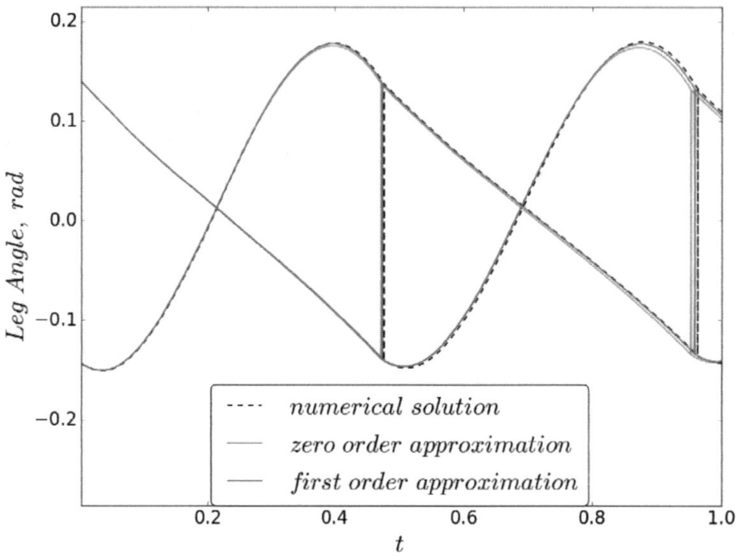

Figure 4. A numerical solution of equation (2) (dotted line) and the asymptotic approximation (solid lines).

Conclusions

We considered a hybrid model of a biped passive walker with two independent small parameters, the ramp slope, α, and the ratio of the mass of the leg to the mass at the hip joint, β. Using the regular asymptotic expansion, we obtained an approximating system and showed that its stable periodic solution is in a good agreement with the numerical solution of the model for small α and β. The asymptotic solution including the first correction to the zero order approximation improves the agreement between the solutions. It should be noted that the model of the biped walker naturally reproduces a stable gait on small slopes (small α) for a relatively wide range of possible ratios between leg and body masses, β. The asymptotic approximation that we obtained by the perturbation method works well when both α and β are small.

Acknowledgements

SY and VS were supported in part by the Russian Foundation for Basic Research through grant 16-41-630524 and by the Ministry of Education and Science of the Russian Federation under the Competitiveness Enhancement Program of Samara University (2013-2020). DR acknowledges the support of NSF through grant DMS-1413223.

References

[1] McGeer T 1990 Passive Dynamic Walking. *International Journal of Robotics Research*, **9(2)**, 62-82
[2] Spong M W and Bhatia G 2003 Further results on control of the compass gait biped. *International Conference on Intelligent Robots and Systems*, 1933-1938
[3] Goswami A, Espiau B and Keramane A 1996 Limit cycles and their stability in a passive bipedal gait. *IEEE Conference on Robotics and Automation*, 246-251
[4] McGeer T 1993 Dynamics and control of bipedal locomotion. *Journal of Theoretical Biology*, **166(3)**, 277-314
[5] Arthur D. Kuo 2002 Energetics of actively Powered locomotion using the simplest walking model. *Journal of Biomechanical Engineering* **124**, 113-120

[6] Collins S, Ruina A, Tedrake R and Wisse M 2005 Efficient bipedal robots based on passive dynamic walkers *Science* **307**, 1082-1085

[7] Garcia M, Chatterjee A, Ruina A and Coleman M.J 1998 The Simplest Walking Model: Stability, Complexity, and Scaling. *ASME Journal of Biomechanical Engineering*, **120**, 281-288.

[8] Goswami A, Espiau B and Keramane A 1997 Limit cycles in a passive compass gait biped and passivity-mimicking control laws. *Journal of Autonomous Robots*, **4(3)**,

[9] Shah N and Yeolekar M 2015 Influence of Slope Angle on the Walking of Passive Dynamic Biped Robot. *Applied Mathematics*, **6(3)**: 456-465

[10] Goswami A, Thuliot B, Espiau B 1996 Compass-Like Biped Robot Part I: Stability and Bifurcation of Passive Gaits.*INRIA* RR-2996,